POWER MATH

POWER MATH

30 Powerful Principles for Building Math & Numeracy Skills

BRANDON ROYAL

JAICO PUBLISHING HOUSE
Ahmedabad Bangalore Chennai
Delhi Hyderabad Kolkata Mumbai

Published by Jaico Publishing House
A-2 Jash Chambers, 7-A Sir Phirozshah Mehta Road
Fort, Mumbai - 400 001
jaicopub@jaicobooks.com
www.jaicobooks.com

Published in arrangement with
Maven Publishing
4520 Manilla Road S.E.
Calgary, Alberta, Canada T2G 4B7

To be sold only in India, Bangladesh, Bhutan,
Pakistan, Nepal, Sri Lanka and the Maldives.

POWER MATH
ISBN 978-81-8495-730-3

First Jaico Impression: 2015
Third Jaico Impression: 2017

Printed by
Trinity Academy For Corporate Training Limited, Mumbai

The beauty of math
is that things always add up.
Success in math is 70 percent
hard work, 30 percent skill,
and 10 percent luck.

—Anonymous

Contents

Introduction

The primary goal of this book is to help readers develop those thinking skills that are essential for mastering basic math, and to do so in the shortest possible time frame. Such thinking skills are closely aligned with numeracy skills. The person who is numerate grasps the "how" and "why" of problem solving. He or she has a dexterity with numbers and is much more likely to approach problem solving in a conceptual way with an understanding of the key math principles involved. Numeracy is more than performing calculations accurately; it combines the science of math with the art of numbers.

Power Math provides a compilation of the most useful academic and real-life math concepts. The secret recipe used in preparing material for this book is a blend of classic problems and relevant tips. Classic problems are recurring, value-added problems which give you, the reader, maximum return for your time and effort spent. Tips serve as math principles; they represent the themes that bind categories of related problems. In addition to presenting 120 classic problems and introducing 30 tips, this book lends a three-tier chili rating system to help readers gauge the difficulty level of problems. A single chili indicates that the difficulty level of a given problem is "mild," two chilies spell "hot," and three chilies signal "very hot."

Good math, like good cuisine, starts with good ingredients.

Chapter 1: Basic Numeracy Ingredients highlights the importance of reviewing the basic building blocks of math, which lie at the foundation of superior problem-solving skills. With these skills come the knowledge and confidence to easily answer the following questions: What are the five different percentage formulas as they relate to percentage increase and decrease? What pitfalls exist in working with ratios and proportions? Does the order of mathematical operations really matter? How do we know if two variables vary directly or indirectly or proportionately or disproportionately? Why is markup always larger than margin in any given scenario?

Good math, like good cuisine, often comes with a recipe.

Chapter 2: Wonderful Math Recipes highlights the importance of knowing the "best" approach. This may involve using shortcuts when solving math problems of a similar category or type. The "group formula" for solving overlap scenarios, the "nine box" table for handling matrix problems, the "barrel method" for solving mixture problems, and the "weighted average" formula for tackling weighted average problems are all examples of tools and techniques that greatly aid problem solving.

Good math, like good cuisine, is practical.

Chapter 3: Favorite Numeracy Dishes highlights math in action. With a focus on business and commerce, the topics addressed include markup versus margin, cost-price-volume-profit analysis, break-even point, efficiency, and distribution and allocation scenarios.

Good math, like good cuisine, has good presentation.

Chapter 4: Special Math Garnishments highlights math from a graphical perspective. Charts and graphs are "math pictures" insofar as they present visual representations of numerical data. Learning how to read graphical information is partly a function of understanding the strength of each type of graph or chart in terms of its use as a presentation tool—line graph, pie chart, or bar chart—and how its use likely ties to the type of data we are trying to summarize. A brief exploration of statistics helps us understand how two variables may be related, why the term "average" can mean different things, and what it means if a finding is considered to be statistically significant.

The primary target audience for this book is the everyday student: the high school, college, or university student. Students studying for the math or quantitative sections of College Board tests and standardized entrance exams—including the SAT, ACT, GRE, and GMAT—will find this material especially useful. Given the pervasive nature of math and numeracy skills, this book is also suitable for any person seeking an on-the-job refresher course, particularly professionals pursuing work in banking and management consultancy, accounting, law, and government.

How is *Power Math* uniquely positioned within the marketplace of math books? Teachers and employers frequently complain that students or employees are weak in the basics of math or lacking in numeracy skills. Unfortunately, there exists neither a definable, concrete list as to what constitutes the basics nor an efficient method to acquire such skills. This book helps address this deficiency. It provides an answer to the key question: "What are some of the key math topics and types of problems that a person should know in order to master the basics of math?"

With reference to the material in this book, the term "basics" refers to simple but insightful arithmetic and algebra. More poetically, mastering the basics refers to the point at which a person starts to "dance with math." A rudimentary level of math knowledge is required to solve the majority of problems in this book. Apart from the ability to do arithmetic, including the ability to add, subtract, multiply, and divide, a reader may be called upon to perform basic algebra, including solving for a single variable. Even if a reader can't recall the formal techniques used to solve basic math problems, the explanations provided for those problems should provide a helpful bridge.

Although this material was specifically designed to help individuals develop a numerical mindset in the shortest possible time frame, the greatest benefit to be gained from this book is likely the inspiration to want to continue learning—regardless of the skill area. The satisfaction gained from those "aha" learning moments, coupled with a sense of confidence from taking "ownership" in the material at hand, is a wondrous yet real by-product of the learning process.

Let's get started.

Quiz

Try these ten basic, but often tricky, math problems. Mark each statement as being either true or false. Answers are found on pages 273–274.

1. If the ratio of females to males at a business conference is 1:2, then the percentage of people who are female at this conference is 50%. ❑ True ❑ False

2. For a given product, markup is always a smaller percentage than margin. ❑ True ❑ False

3. A couple charged $132 on their credit card to pay for a meal while dining out. This $132 figure included a 20% tip which was paid on top of the price of the meal, which already included a sales tax of 10%. The actual price of the meal before tax and tip was $92.40. ❑ True ❑ False

4. Ratios are useful tools that tell us something about actual size or value. ❑ True ❑ False

5. Multiplying a number by 1.2 is the same as dividing that same number by 0.8. ❑ True ❑ False

6. Break-even occurs exactly where profit equals total fixed costs. ❑ True ❑ False

7. A store item that has been discounted first by 20% and then by 30% is now selling at 50% of its original price. ❑ True ❑ False

8. At a summer party attended by exactly 100 persons, invitees were asked to consider making a small donation to one of Charity A or Charity B or to both charities. Sixty persons donated only to Charity A and 35 persons donated only to Charity B, while 20 persons donated to neither Charity A nor Charity B. Based on this information, we can conclude that 25 persons donated to both Charity A and Charity B. ❑ True ❑ False

9. If product A is selling for 20% more than product B, then the ratio of the selling price of product A to the selling price of product B is 100% to 80%. ❑ True ❑ False

10. Data with a high standard deviation is "bunched." Data with a low standard deviation is more "spread out." ❑ True ❑ False

Chapter 1

Basic Numeracy Ingredients

Math class is tough.

—Barbie's original voice chip by Mattel

Percentages

☞ Tip #1: **A percentage increase is not the same as a percentage of an original number.**

Snapshot

A percentage is a number expressed in relation to 100. The word "percentage" means "hundredths." The percentage sign (%) represents the fraction $\frac{1}{100}$ or its decimal equivalent, 0.01. A percentage is a number that relates two quantities that are usually referred to as the whole and the part.

Below is the most basic percentage formula, as well as two variants:

$$\text{Whole} \times \text{Percent} = \text{Part} \qquad 100 \times 25\% = 25$$

$$\frac{\text{Part}}{\text{Whole}} = \text{Percent} \qquad \frac{25}{100} = 25\%$$

$$\frac{\text{Part}}{\text{Percent}} = \text{Whole} \qquad \frac{25}{25\%} = 100$$

The following problems highlight the five "classic" percentage formulas used to calculate percentage increase and decrease:

(1) Percentage of an original number

A stock is \$100 at the beginning of the year and \$120 at the end of the year. The ending value is what percentage of the beginning value?

$$\frac{\text{New}}{\text{Old}} \qquad \frac{\$120}{\$100} = 120\%$$

(2) Percentage increase

A stock is $100 at the beginning of the year and $120 at the end of the year. By what percentage has its value increased?

$$\frac{\text{New} - \text{Old}}{\text{Old}} \qquad \frac{\$120 - \$100}{\$100} = \frac{\$20}{\$100} = 20\%$$

(3) Percentage decrease

A stock is $100 at the beginning of the year and $80 at the end of the year. By what percentage has its value decreased?

$$\frac{\text{Old} - \text{New}}{\text{Old}} \qquad \frac{\$100 - \$80}{\$100} = \frac{\$20}{\$100} = 20\%$$

(4) Percentage decrease to return to an original number

A stock is $100 at the beginning of the year and $120 at the end of the year. By what percentage would the price of the stock have to be decreased in order to return it to its original price?

$$\frac{\text{New} - \text{Old}}{\text{New}} \qquad \frac{\$120 - \$100}{\$120} = \frac{\$20}{\$120} = 16.67\%$$

(5) Percentage increase to return to an original number

A stock is $100 at the beginning of the year and $80 at the end of the year. By what percentage would the price of the stock have to be increased in order to return it to its original price?

$$\frac{\text{Old} - \text{New}}{\text{New}} \qquad \frac{\$100 - \$80}{\$80} = \frac{\$20}{\$80} = 25\%$$

Each of the following five problems can be solved using one of the five classic percentage formulas. Answers begin on page 180.

1. Antique Stamp ( )

The trading value of an antique stamp rose in price from $100 to $150. The current price is what percentage of the original price?

A) 20%
B) 25%
C) $33\frac{1}{3}\%$
D) 50%
E) 150%

2. Rise ( )

The trading value of an antique stamp rose in price from $100 to $150. What was the percentage increase in price?

A) 20%
B) 25%
C) $33\frac{1}{3}\%$
D) 50%
E) 150%

3. Fall ( )

The trading value of an antique stamp declined in price from $100 to $50. What was the percentage decrease in price?

A) 20%
B) 25%
C) $33\frac{1}{3}\%$
D) 50%
E) 150%

4. Devalue ( )

The trading value of an antique stamp rose in price from \$100 to \$125. By what percentage would the price of the stamp have to be decreased in order to return it to its former price?

A) 20%
B) 25%
C) $33\frac{1}{3}\%$
D) 50%
E) 150%

5. Revalue ( )

The trading value of an antique stamp declined in price from \$100 to \$75. By what percentage would the price of the stamp have to be increased in order to return it to its former price?

A) 20%
B) 25%
C) $33\frac{1}{3}\%$
D) 50%
E) 150%

6. Twist ( )

The price of a stock ended the year four times higher than it was at the beginning of the year. By what percentage had the stock increased in price?

A) 75%
B) 125%
C) 200%
D) 300%
E) 400%

7. Microbrewery ( )

Over the course of a year, a certain microbrewery increased its output by 70%. At the same time, it decreased its total working hours by 20%. By what percentage did this factory increase its output per hour?

A) 50%
B) 90%
C) 112.5%
D) 210%
E) 212.5%

8. Gardener ( )

A gardener increased the length of his rectangle-shaped garden by increasing its length by 40% and decreasing its width by 20%. The area of the new garden

A) has increased by 20%
B) has increased by 12%
C) has increased by 8%
D) is exactly the same as the old area
E) cannot be expressed in percentage terms without knowing the dimensions of the original garden

☞ Tip #2: You can't add (or subtract) the percentages of different wholes.

Snapshot

There is a story about a man and his fish farm. The man bought the farm, paid for an initial stock of fish, and hired a manager. At the end of his first year of operation, the new owner was told by the manager that there had been a 100 percent increase in the number of fish on the farm. At the end of the second year of operation, the manager told the owner that, due to harvesting, the number of fish on the farm had decreased by 50 percent. Upon hearing this, the owner thought to himself: "I had a 100 percent increase in the number of fish during the first year and then a 50 percent decrease in the number of fish during the second year. I must still have 50 percent more fish on the farm then when I first started."

Of course, our dear owner did not have 50 percent more fish because we can't compare (in this case, subtract) the percentages of different wholes. The owner has exactly the same number of fish on the farm as he did when he first started. Let's use some simple numbers to prove this. Say the owner originally bought a stock of 100 fish. During the first year, the number of fish increased 100 percent or from 100 fish to 200 fish. During the second year, the number of fish decreased 50 percent or from 200 fish to 100 fish. There are still 100 fish on the farm.

In this example, the wholes involve 100 fish and 200 fish. These are the number of fish at the beginning of the first year and second year, respectively. There was an actual increase of 100 fish and a decrease of 100 fish. In percentage terms, however, the increase represents an increase of 100 percent while the decrease represents a decrease of only 50 percent. This occurs because the percentage increase is based on a smaller base (that is, 100) while the percentage decrease is based on a larger base (that is, 200). Simply put, 100 percent of 100 is the same as 50 percent of 200.

Compounding of interest presents an everyday example where we cannot simply add percentages together because the base amount (that is, principal) increases due to compounding. For example, leaving $1,000 in a savings account that pays 10% annual interest would, at the end of two years, yield more than $200 in interest (that is, 20% of $1,000). The reason for this is that in the first year, we multiply 10% by the base amount of $1,000. In the second year, we multiply 10% by the base amount of $1,100. The wholes $1,000 and $1,100 are not the same.

Total interest equals:

	Principal × Rate = Interest
First Year:	$1,000 × 10% = $100
Second Year:	$1,100 × 10% = $110
	$210

An alternative, quicker calculation, unfolds as follows:

	Principal × $Rate_1$ × $Rate_2$ = Total
Years 1 & 2:	($1,000 × 110%) × 110% = $1,210
Less Principal:	$1,000
	$210

9. Train ( )

A passenger train increases its speed by 25% and then increases its speed by 20%. What is its overall increase in speed?

A) 25%
B) 45%
C) 50%
D) 145%
E) 150%

10. Broker

A broker invested her own money in the stock market. During the first year, she increased her stock market wealth by 50 percent. In the second year, largely as a result of a slump in the stock market, she suffered a 30 percent decrease in the value of her stock investments. What was the net increase or decrease on her overall stock investment wealth by the end of the second year?

A) –5%
B) 5%
C) 15%
D) 20%
E) 80%

11. Double Discount

A discount of 20% on an order of goods followed by a discount of 10% amounts to

A) the same as one 15% discount
B) the same as one 22% discount
C) the same as one 25% discount
D) the same as one 28% discount
E) the same as one 30% discount

12. Net Effect #1

If the price of an item is decreased by 10% and then increased by 10%, the net effect on the price of the item is

A) no change
B) an increase of 1%
C) a decrease of 1%
D) a decrease of 99%
E) variable depending on the actual price involved

13. Net Effect #2 ( )

If the price of an item is increased by 20% and then decreased by 20%, the net effect on the price of the item is

A) no change
B) an increase of 4%
C) a decrease of 4%
D) a decrease of 96%
E) variable depending on the actual price involved

14. Silver ( )

The price of silver rose by 100% and then fell by 50%. Compared to the original price, the final price is

A) 50% greater
B) 25% greater
C) thc same
D) 25% less
E) 50% less

15. Growth ( )

A business is growing at 10% a year. How many years will it take for the business to double in size?

A) less than 10 years
B) exactly 10 years
C) greater than 10 years
D) either less than or greater than 10 years
E) it depends on the size of the company

☞ Tip #3: **Percentages cannot be compared directly to numbers because percentages are relative measures while numbers are actual values.**

Snapshot

We generally cannot compare percentages to numbers unless we know the exact numbers represented by those percentages. Percentages, like decimals, fractions, and ratios, are relative measures; actual numbers provide absolute measures.

Let's presume that in a given company 10% of the employees in department A are salespersons, whereas 20% of the employees in department B are salespersons. Does department B have more salespersons than department A? The answer, of course, is that we cannot tell. It could be that both departments have the same number of total employees, in which case department B would definitely have more salespersons than department A (see scenario 1). But there could be many more employees working in department A than in department B. In this case, the number of salespersons working in department A could well be greater than the number of salespersons working in department B (see scenario 2).

Scenario 1

	Dept A	Dept B
Number of total employees	100	100
% of employees who are salespersons	10%	20%
Number of salespersons	10	20

Scenario 2

	Dept A	Dept B
Number of total employees	100	40
% of employees who are salespersons	10%	20%
Number of salespersons	10	8

16. Medical School ( )

A student activist group has released a report that suggests that female students have more difficulties in earning admittance to medical school. The facts in the report speak for themselves: 75% of all students in medical school are male, but fewer than half of all female applicants reach their goal of being admitted to medical school in any given year.

Which of the following data would be most helpful in evaluating the argument above?

A) The proportion of all admissions officers who are male versus female.

B) The percentage of eligible female students admitted to medical school who accept a particular school's offer of admission, and the percentage of eligible male students admitted to medical school who also accept that particular school's offer of admission.

C) The percentage of eligible female students admitted to medical school, and the percentage of eligible male students admitted to medical school.

D) Comparative records on acceptance rates of male and female applicants in other graduate programs such as business, law, and the sciences.

E) The dropout rate for female and male students while in medical school.

17. White-Collar Crime ( )

Sociologist Dr. Noah is inconsistent in what he has to say about the causes of white-collar crime. He claims that unemployment, often aggravated by alcoholism, accounts for at least 50% of the increase in the rate of white-collar crime over the past year. How can he reconcile this with his opinion, expressed earlier in the very same lecture, that the loneliness that results from overwork is at the root of most of the white-collar crime that occurred last year?

The author's criticism of Dr. Noah's comments is

A) Warranted, because it cannot be true that most white-collar crime is caused by loneliness if such crime is caused by economic factors.

B) Warranted, because Dr. Noah fails to specify the number of white-collar crimes that occurred ten years ago.

C) Unwarranted, because unemployment and loneliness can both be factors in the same white-collar crime.

D) Unwarranted, because the author fails to mention that severe depression unrelated to unemployment or loneliness can precipitate white-collar crimes.

E) Unwarranted, because there is a difference between the total number of white-collar crimes last year and the number represented by the rate of increase in those crimes last year.

18. Fly Fishing

According to *Angler's Digest* online magazine, the percentage of anglers in Northern Ontario who are using fly fishing, as opposed to bait fishing, has increased 10% over the last 20 years. If the statements above are true, all of the following statements concerning sport fishing in Northern Ontario during the past two decades could also be true EXCEPT

A) The actual number of people fly fishing in Northern Ontario has increased while the overall number of people fishing for sport in that region has decreased.

B) The actual number of people fly fishing in Northern Ontario has increased while the overall number of people fishing for sport in that region has remained the same.

C) The actual number of people fly fishing in Northern Ontario has increased while the overall number of people fishing for sport in that region has also increased.

D) The actual number of people fly fishing in Northern Ontario has decreased while the overall number of people fishing for sport in that region has also decreased.

E) The actual number of people fly fishing in Northern Ontario has remained the same while the overall number of people fishing for sport in that region has increased.

☞ Tip #4: **Percentages (or ratios) are necessary tools for drawing conclusions about qualitative measures, such as likability, preferability, danger, or safety.**

Snapshot

Is train travel becoming more dangerous? Even if, in a given year, there are more train crashes and regrettably more injuries or deaths compared with a previous year, this does not necessarily mean that train travel is becoming more dangerous. We have to examine how many people are riding on trains (or, more precisely, how many injuries/deaths per train mile logged). If many more people are taking trains, it is likely that the percentage or ratio of train related injuries/deaths (to total passengers) is decreasing. In these circumstances, train travel would be considered safer, not more dangerous. Ratios (or fractions, decimals, or percents) are necessary tools for drawing conclusions about qualitative measures such as likability, preferability, danger, or safety.

19. Lottery (🌶)

Annual Incomes	Number of ticket purchases
Less than $20,000	10,000
$20,001 – $50,000	7,000
$50,001 – $100,000	4,000
Over $100,000	800

The data above, based on a recent report prepared by Lotto Corp., indicates that the number of people who spend $1 to buy a lottery ticket is greater for people with lower levels of annual income than it is for people with higher levels of annual income. Assuming that this survey was representative of the buying patterns of individuals within each income category, are people who make more money less likely to buy lottery tickets?

A) Yes
B) No
C) Can't Tell

20. Iraq ()

During the first three years of the Second Gulf War, 2,300 members of the United States Armed Forces died fighting in Iraq. During this same three-year period, some 100,000 civilians died in the United States from automobile-related accidents. These figures reveal a sad fact: It was more dangerous for a civilian to drive on the streets and highways of America than it was to be dressed in uniform and serving in Iraq.

Which of the following calculations, during this same three-year period, would reveal most clearly the absurdity of the conclusion drawn above?

A) Comparing deaths caused by combat in the U.S. armed forces in Iraq to gun-related deaths in the domestic U.S.

B) Comparing the number of deaths among U.S. civilians that resulted from driving at high speeds to the total number of deaths among members of the armed forces.

C) Factoring out deaths caused by accidents during service in the armed forces in Iraq as well as those due to traffic deaths in the U.S. which were not considered to be the fault of those individuals.

D) Comparing death rates per thousand members of each group rather than total number of deaths.

E) Factoring into consideration the small percentage of time that the average driver in the U.S. drives his or her car each day compared with the full-time status of members of the armed forces in Iraq.

Tip #5: ***Growing pies present a classic case where the actual dollar value of a given item increases over time, yet this same item decreases in percentage terms relative to a larger pie.***

Snapshot

Growing or shrinking pies involve situations where the distinction between numbers and percents can become blurred.

Growing Pie—You may spend more money on personal phone calls this year compared with last year, but this does not mean that the percentage of your disposable income that you spend on phone calls has gone up. It is quite possible, for example, that you have enjoyed a significant increase in your disposable income. Therefore, although the actual money that you spend on personal phone calls has gone up, the percentage of your disposable income that you spend on phone calls has gone down. In short, a growing pie often results in an increase in dollar expenditures despite a corresponding decrease in the percentage of money being spent on a particular item or thing.

Shrinking Pie—You may spend less money dining out than you used to, but this does not mean that you spend a smaller percentage of your money dining out. What if you have a lot less disposable income these days? This would imply that a lesser (or even equal) amount of money spent dining out would mean that a greater portion or percentage of your disposable income is being spent dining out. In short, a shrinking pie often results in a decrease in dollar expenditures despite a corresponding increase in the percentage of money being spent on a particular item or thing.

21. Military Expenditures ( )

Proponents of greater military spending for Reno Republic argue that the portion of their national budget devoted to military programs has been declining steadily for a number of years, largely because of rising infrastructure development costs. Yet groups opposed to increasing military expenditure point out that the Reno Republic military budget, even when measured in constant, or inflation-adjusted, dollars, has increased every year for the past two decades.

Which of the following, if true, best resolves the apparent contradiction presented above?

A) Countries in the same region as the Reno Republic have increased the annual portion of their national budgets devoted to military spending.

B) The total national budget has increased at a faster rate than military expenditure has increased.

C) The advocates of greater military spending have overestimated the amount needed for adequate defense of the country.

D) Military expenditures have risen at a rate higher than the rate of inflation.

E) Total military expenditures have increased faster than the national budget has increased.

☞ Tip #6: **Consider the following statement: "Person A spends a greater portion of his or her disposable income on beer than on coffee as compared to person B." This does not necessarily mean that person A spends more money in total on beer and coffee than does person B. It also does not mean that person A, individually, spends more money on beer than he or she does on coffee.**

Snapshot

To preview the concept tested in the upcoming problem, titled *Fiction Books,* ponder the following statement:

> The percentage of white wine to red wine bottled in Germany is greater than the percentage of white wine to red wine bottled in France.

Does this mean that there is more white wine bottled in Germany than white wine bottled in France? Answer—No. Why? Because the size of the wine industry in France is so much bigger than that of the wine industry in Germany. The French "pie" is much bigger than the German "pie," five times bigger, in fact.

Does the original statement mean that Germany bottles more white wine than it does red wine or that France bottles more white wine than it does red wine? The answer is no on both accounts.

Let's draw upon some hypothetical but plausible numbers in line with statistics from the wine industry. France produces five times as much wine as Germany does. Of all the wine produced by Germany, 75% is white and 25% is red. Of all the wine produced by France, 25% is white and 75% is red. Even though the percentage of white wine to red wine bottled in Germany is greater than the percentage of white wine to red wine bottled in France, France still produces more white wine than Germany does. Here are some simple numbers to prove this point. Say that France produces 500 bottles of wine each year and Germany produces 100 bottles of wine each year. This means that France produces 125 bottles of white wine whereas Germany produces 75 bottles of white wine.

French Wine Production 500 bottles		German Wine Production 100 bottles	
(White Wine) 125 bottles	(Red Wine) 375 bottles	(White Wine) 75 bottles	(Red Wine) 25 bottles

22. Fiction Books ( )

A major publishing conglomerate released a survey concerning the relationship between a household's education level and the kind of books found in its library. Specifically, members of households with higher levels of education had more books in their libraries. The survey also indicated that compared with members of households with lower levels of education, members of households with higher levels of education had a greater percentage of fiction versus nonfiction books.

Which of the following can be properly inferred from the survey results cited above?

A) People with the highest levels of education buy more fiction books than nonfiction books.

B) Members of households with higher levels of education have more fiction books than nonfiction books.

C) Members of households with lower levels of education have more nonfiction books than fiction books.

D) Members of households with lower levels of education have more nonfiction books than do members of households with higher levels of education.

E) Members of households with higher levels of education have more fiction books than do members of households with lower levels of education.

Ratios and Proportions

☞ Tip #7: **Ratios tell us nothing about actual size or value.**

Snapshot

Ratios

A ratio is a comparison. There are two ways to write a ratio. The first uses the word "to." The second uses a colon, which consists of a pair of vertical dots. Both of the representations below have identical meanings.

7 to 8 or 7:8

We can and should always reduce a ratio. So 4:8 is written 1:2. Ratios should be expressed using integers, not fractions or decimals.

Example $\frac{1}{2}:2 \rightarrow 1:4$ $1:1.5 \rightarrow 2:3$

It is important to remember that ratios, like percentages, decimals, and fractions, are relative measures. They are not "actual numbers" that otherwise provide absolute values. The following are examples of relative measures:

- the percentage of your disposable income that you spent on vacations last year
- the fraction of your free time that you watch TV each week
- the ratio of your cash savings to your personal debt at this very moment

Now take, for example, the following absolute measures:

- the dollars you actually spent on vacations last year
- the number of hours you watch TV each week
- the amount of money you currently have as savings and personal debt

The beauty of ratios (as well as percentages, fractions, and decimals) is that they allow us to compare things of unequal size. Ratios are particularly useful for making relative comparisons, especially when two items differ greatly in size.

Consider these two statements: "Singapore is the most densely populated developed country in the world and Australia is the least densely populated developed country in the world. The population density of Singapore is 7,300 people per square kilometer while the population density of Australia is 3 people per square kilometer."

Given that we are using ratios to express the number of people per square kilometer, it is not necessary to worry about the actual populations or land areas of Singapore and Australia.

The "weakness" of ratios (and percentages, fractions, and decimals) is that they obscure size. This distinction may be critical in a given situation. Case in point: Two businesses may have identical—2 to 1—ratios in terms of the amount of total assets compared to total liabilities. With respect to this ratio anyway, they are on relative par. On an absolute scale, however, their financial realities might be starkly different. One of these businesses might be Ken's Kiosk, for which the owner has $2,000 in total assets and $1,000 in total debt. The other business might be Super Corp, which has 2 billion dollars in total assets and 1 billion dollars in total debt. One business has $1,000 in net assets while the other has $1 billion in net assets.

Proportions

In the same way that a ratio is a comparison between two (or more) numbers, a proportion is a comparison between two (or more) ratios. Specifically, a proportion is an equation that shows that two ratios are equal. Equivalent or equal fractions form a proportion. There are two ways to write a proportion. The first uses four dots: "::". The second uses an "=" sign to express a proportion as an equation. Thus, the following are equivalent:

Examples $\frac{a}{b} :: \frac{c}{d}$ $\frac{1}{2} = \frac{50}{100}$

In a proportion, cross products are equal:

Example $\frac{a}{b} :: \frac{c}{d}$ $a \times d = b \times c$

Example $\frac{1}{2} = \frac{50}{100}$ $1 \times 100 = 2 \times 50$

Solving proportions as equations:

$$\frac{3}{5} = \frac{x}{50}$$

$$3 \times 50 = 5x$$

$$150 = 5x$$

$$5x = 150$$

$$x = \frac{150}{5}$$

$$x = 30$$

There are four ways to structure proportions:

$\frac{\text{Part}}{\text{Whole}} :: \frac{\text{Part}}{\text{Whole}}$ Example $\frac{3}{10} :: \frac{300}{1000}$

or

$\frac{\text{Whole}}{\text{Part}} :: \frac{\text{Whole}}{\text{Part}}$ Example $\frac{10}{3} :: \frac{1000}{300}$

or

$\frac{\text{Part}}{\text{Part}} :: \frac{\text{Whole}}{\text{Whole}}$ Example $\frac{3}{300} :: \frac{10}{1000}$

or

$\frac{\text{Whole}}{\text{Whole}} :: \frac{\text{Part}}{\text{Part}}$ Example $\frac{10}{1000} :: \frac{3}{300}$

23. WAA ( )

The World Automobile Association (WAA) publishes a list of the "Best and Worst Drivers of the World," ranking the drivers of every nation according to the number of traffic deaths per mile driven in that country.

Each of the following, if true, would by itself provide a logical objection to using the WAA's ranking as a representation of the quality of drivers in each nation EXCEPT

A) Some countries contain hundreds of thousands of miles of road while other countries contain relatively few miles of road.

B) The roads in some countries are in bad repair and are therefore more dangerous than roads in other countries.

C) The average driver in industrialized countries can afford to maintain his or her car in better condition than can the average driver in less developed countries.

D) Minor accidents that would cause little injury in many countries are often fatal when they occur in extremely mountainous countries.

E) Because of differences in national economies, the average car in some countries contains many more passengers than does the average car in other countries.

☞ **Tip #8:** **You cannot add or subtract a number from a ratio unless you know the exact numbers that make up the ratio. However, you can always multiply or divide both sides of a ratio by a given number.**

Snapshot

Ponder the following: "The ratio of white goldfish to orange goldfish in a tank is 3:10. If two white goldfish are added to the tank, what is the new ratio of white to orange goldfish?"

If your instinct is to blurt out "50 percent," you're not alone. After all, if we add 2 and 3 to get 5, we can create a new ratio of 5:10. This simplifies to 1:2 or 50 percent. You might, on the other hand, be second-guessing yourself, wondering whether the answer is 5:12. After all, 2 + 3 = 5 and 2 + 10 = 12. This answer, however, is clearly flawed because we could never just mathematically add two white goldfish to the number of orange gold fish.

The answer is that we can't tell. It could be that the ratio is indeed 5:10 or 50 percent. But this assumes that there are exactly 3 white goldfish and 10 orange goldfish to begin with. What if there are 3,000 white goldfish and 10,000 orange goldfish? Then the ratio of 3 to 10 white to orange goldfish will hardly change when two white goldfish are added to the tank.

The takeaway here is that we can only add or subtract from a ratio when we know the exact numbers that make up the ratio.

24. Lemonade ( )

A lemonade mixture that is 1 part lemonade concentrate to 2 parts water is added to another mixture that is 2 parts lemonade concentrate to 3 parts water. What is the ratio of lemonade concentrate to water in the resulting mixture?

A) 3:5
B) 3:8
C) 5:3
D) 8:3
E) Cannot be determined

Tip #9: A part-to-part ratio is not the same thing as a part-to-whole ratio. If the ratio of female to male attendees at a conference is 1:2, this does not mean that 50 percent of the conference attendees are female.

Snapshot

If the ratio of women to men at a conference is 1 to 2, what percentage of the conference is female? The answer is not 50%, but instead $33\frac{1}{3}\%$. That is, the ratio of women to total attendees is 1:3 or $33\frac{1}{3}\%$. It is easy to confuse part-to-part ratios with part-to-whole ratios. The ratio of women to men is a part-to-part ratio; the ratio of women to total attendees is a part-to-whole ratio. Just substitute some numbers to make everything clear. Say there are 5 females and 10 males at the conference. What percentage of the conference attendees are female? Answer: $\frac{5}{15} = \frac{1}{3} = 33\frac{1}{3}\%$.

Ratio of women to men at the conference	Ratio of women to total attend-ees at the conference	Fraction of women to total attend-ees at the conference	Percentage of women to total attend-ees at the conference
1 to 2	1 to 3	$\frac{1}{3}$	$33\frac{1}{3}\%$

Encore: The earth's surface is approximately 1 part land to 3 parts water. What percentage of the earth's surface is land? It's easy to think the answer is one-third or $33\frac{1}{3}\%$. But, in fact, it's one part land to 4 parts (land plus water). The answer is $\frac{1}{4}$ or 25%.

25. Eggs (🌶)

In a box of one dozen eggs, the ratio of broken eggs to unbroken eggs could be any of the following EXCEPT

A) 1:1
B) 1:2
C) 1:3
D) 1:4
E) 1:5

26. Elk Herd (🌶)

If the ratio of male elk to female elk in a large herd is 16:9, then what fraction of the herd is male?

A) $\frac{16}{9}$ B) $\frac{16}{25}$ C) $\frac{9}{16}$ D) $\frac{9}{25}$ E) $\frac{1}{4}$

27. Rum & Coke ( )

A glass holding 6 ounces of a rum drink that is 1 part rum to 2 parts Coke is added to a jug holding 32 ounces of a rum drink that is 1 part rum to 3 parts Coke. What is the ratio of rum to Coke in the resulting mixture?

A) 2:5
B) 5:14
C) 3:5
D) 4:7
E) 14:5

28. Toledo Prep ( )

At Toledo Prep School, the faculty-to-student ratio is 1:7. If 3 of every 7 students are female and one-fifth of the faculty is female, what fraction of the combined students and faculty are female?

A) 30%
B) 32%
C) 40%
D) 42%
E) 57%

29. Deluxe ( )

At Deluxe paint store, fuchsia paint is made by mixing 5 parts of red paint with 3 parts of blue paint. Mauve paint is made by mixing 3 parts of red paint with 5 parts blue paint. How many liters of blue paint must be added to 24 liters of fuchsia to change it to mauve paint?

A) 9
B) 12
C) 15
D) 16
E) 18

Tip #10: Proportions, in the form of $\frac{A}{B} = \frac{C}{D}$, are among the simplest but most versatile math problem-solving tools.

Snapshot

If 1 kilometer is equal to 0.6 miles, then how many kilometers are equal to 12 miles?

Ratio Structure:

$$\frac{\text{Part}}{\text{Part}} :: \frac{\text{Whole}}{\text{Whole}}$$

Applied to this problem:

$$\frac{\text{kilometers}}{\text{miles}} :: \frac{\text{kilometers}}{\text{miles}}$$

Calculation:

$$\frac{1\text{ kilometer}}{0.6\text{ miles}} = \frac{x\text{ kilometers}}{12\text{ miles}}$$

$$1(12) = x(0.6)$$

$$x(0.6) = 1(12)$$

$$\frac{1}{\cancel{0.6}} \times x(\cancel{0.6}) = \frac{1}{0.6} \times 1(12)$$

$$x = \frac{12}{0.6}$$

$$x = 20\text{ kilometers}$$

30. Exchange ( )

If one Philippine peso can be exchanged for 0.15 Hong Kong dollars, then how many pesos will a person receive from exchanging one Hong Kong dollar?

A) 0.15
B) 0.67
C) 1.50
D) 6.67
E) 15.0

31. Sale Time ( )

During a holiday special, a customer buys a pair of shoes for $12 off the regular retail price. If all items were marked 30% off, what was the original price of the shoes?

A) $15.60
B) $28.00
C) $36.00
D) $40.00
E) $52.00

32. Landsdown ( )

In the township of Landsdown, 12% of its residents, or 18,000 people, voted in last month's referendum. Assuming these figures are accurate, how many people reside in Landsdown?

A) 90,000
B) 150,000
C) 180,360
D) 182,160
E) 216,000

33. Shrinkage ( )

The manager of "I Scream for Summer," an outdoor ice cream kiosk, must make an allowance for an average 25% loss of overall weight that occurs when ice cream melts, evaporates, or spoils in hot weather. If this 25% loss figure is accurate, how many ounces of ice cream (on average) are needed to yield a deluxe 24-ounce dish of ice cream?

A) 18
B) 30
C) 32
D) 36
E) 40

Tip #11: If product A is selling for 20% more than product B, then the ratio of the selling price of product A to the selling price of product B is 120% to 100% (not 100% to 80%). If product B is selling for 20% less than product A, then the ratio of the selling price of product A to the selling price of product B is 100% to 80% (not 120% to 100%).

Snapshot

When visualizing percent problems, relate everything to a 100 percent base. If we have a 20% increase, this means that we started at 100% and have moved to 120% ($\frac{120\% - 100\%}{100\%} = \frac{20\%}{100\%} = 20\%$). Note that it is not the same as starting at 80% and moving to 100%. After all, going from 80% to 100% represents an increase of 25%, not 20% ($\frac{100\% - 80\%}{80\%} = \frac{20\%}{80\%} = 25\%$).

Likewise, if something decreases by 20%, we visualize starting at 100% and moving down to 80% ($\frac{100\% - 80\%}{100\%} = \frac{20\%}{100\%} = 20\%$). This is not the same as moving from 120% to 100%. After all, going from 120% to 100% represents a decrease of $16\frac{2}{3}\%$, not 20% ($\frac{120\% - 100\%}{120\%} = \frac{20\%}{120\%} = 16\frac{2}{3}\%$).

34. Product A ( )

Product A sells for 20% more than product B, and product B sells for $40. What is the price of product A?

A) $32.00
B) $40.00
C) $41.67
D) $48.00
E) $50.00

35. Product B ( )

Product A sells for 20% more than product B, and product A sells for $50. What is the price of product B?

A) $32.00
B) $40.00
C) $41.67
D) $48.00
E) $50.00

36. Product C ( )

Product D sells for 20% less than product C, and product D sells for $40. What is the price of product C?

A) $32.00
B) $40.00
C) $41.67
D) $48.00
E) $50.00

37. Product D ( )

Product D sells for 20% less than product C, and product C sells for $50. What is the price of product D?

A) $32.00
B) $40.00
C) $41.67
D) $48.00
E) $50.00

38. Diners ( )

A man charged $264 on his credit card in payment for a family dinner. This $264 figure included a 20% tip which was paid on top of the price of the food, which already included a sales tax of 10%. What was the actual price of the meal before tax and tip?

A) $185
B) $200
C) $203
D) $206
E) $216

39. Investments ( )

A lady sold two small investment properties, A and B, for $24,000 each. If she sold property A for 20% more than she paid for it, and sold property B for 20% less than she paid for it, then, in terms of the net financial effect of these two investments (excluding taxes and expenses), we can conclude that the lady

A) broke even
B) had an overall gain of $1,200
C) had an overall loss of $1,200
D) had an overall gain of $2,000
E) had an overall loss of $2,000

Reciprocals

☞ Tip #12: Dividing by any number produces the same result as multiplying by that number's reciprocal (and vice versa).

Snapshot

Informally speaking, the reciprocal of a number is that number "upside down." Technically, the reciprocal of a number is "another number, which when multiplied by the original number, yields a product of 1."

Ex. $7 \times \frac{1}{7} = 1$ $\qquad$ $\frac{1}{7}$ is the reciprocal of 7 (and vice versa)

Ex. $\frac{1}{2} \times \frac{2}{1} = 1$ $\qquad$ $\frac{2}{1}$ is the reciprocal of $\frac{1}{2}$ (and vice versa)

Ex. $\frac{5}{3} \times \frac{3}{5} = 1$ $\qquad$ $\frac{3}{5}$ is the reciprocal of $\frac{5}{3}$ (and vice versa)

Scenario 1: Multiplying by any number is exactly the same as dividing by that number's reciprocal. For example, multiplying by 2 is the same as dividing by $\frac{1}{2}$.

Ex. $4 \times 2 = 8$ $\qquad$ *or* $\qquad$ $4 \times 2 \rightarrow 4 \div \frac{1}{2} = \frac{\frac{4}{1}}{\frac{1}{2}} = 8$

Scenario 2: Dividing by a number is exactly the same as multiplying by that number's reciprocal. For example, dividing by 2 is the same as multiplying by $\frac{1}{2}$.

Ex. $4 \div 2 = 2$ $\qquad$ *or* $\qquad$ $4 \div 2 \rightarrow 4 \times \frac{1}{2} = 2$

40. Subtle ( )

Dividing 100 by 0.75 will lead to the same mathematical result as multiplying 100 by which number?

A) 0.25
B) 0.75
C) 1.25
D) 1.33
E) 1.75

41. Topsy-Turvy ( )

Multiplying $100 by 1.2 is equal to $120. Instead of multiplying $100 by 1.2 to get $120, we could have also divided the original $100 by ____ to get $120.

A) 0.20
B) 0.80
C) 0.83
D) 1.16
E) 1.20

Order of Operations

☞ Tip #13: The order in which two or more numbers are added or multiplied does not matter. However, the order in which two or more numbers are subtracted or divided does matter.

Snapshot

The commutative law and the associative law are two mathematical laws of operation that are easy to confuse.

Commutative Law

The commutative law states that order doesn't matter. This law holds for addition and multiplication, but not for subtraction or division. For example, when we add (or multiply) two numbers, we can reverse the order in which we add (or multiply) the two numbers and still expect to get the same result. However, when we subtract (or divide) two numbers, we cannot simply reverse the order in which we subtract (or divide) the two numbers and still expect to get the same result.

Addition:

$$a + b = b + a$$

$$2 + 3 = 3 + 2$$

Multiplication:

$$a \times b = b \times a$$

$$2 \times 3 = 3 \times 2$$

BUT NOT:

Subtraction:

$$a - b \neq b - a$$

$$4 - 2 \neq 2 - 4$$

Division:

$$a \div b \neq b \div a$$

$$4 \div 2 \neq 2 \div 4$$

Associative Law

The associative law states that regrouping doesn't matter. This law holds for addition and multiplication, but not for subtraction or division. For example, when we add (or multiply) three numbers, we can add (or multiply) any two numbers before adding or multiplying the remaining number, and still expect to get the same result. However, this treatment will not necessarily hold true for subtraction or division.

Addition:

$$(a + b) + c = a + (b + c)$$

$$(2 + 3) + 4 = 2 + (3 + 4)$$

Multiplication:

$$(a \times b) \times c = a \times (b \times c)$$

$$(2 \times 3) \times 4 = 2 \times (3 \times 4)$$

BUT NOT:

Subtraction:

$$(a - b) - c \neq a - (b - c)$$

$$(8 - 2) - 4 \neq 8 - (2 - 4)$$

Division:

$$(a \div b) \div c \neq a \div (b \div c)$$

$$(8 \div 2) \div 4 \neq 8 \div (2 \div 4)$$

42. Partners ( )

If a, b, c, and d are real numbers, each of the following expressions equals $a \times (b \times c \times d)$ EXCEPT

A) $(d \times c \times b) \times a$
B) $d \times (a \times b \times c)$
C) $(a \times b)(c \times d)$
D) $(a \times d)(b \times c)$
E) $(a \times b)(a \times c)(a \times d)$

43. Bargain ( )

A discount of 10% on an order of goods followed by a discount of 30% amounts to

A) the same as one 33% discount
B) the same as one 40% discount
C) the same as a discount of 30% followed by a discount of 10%
D) less than a discount of 30% followed by a discount of 10%
E) more than a discount of 30% followed by a discount of 10%

44. Inflation ( )

An inflationary increase of 20% on an order of raw materials followed by an inflationary increase of 10% amounts to

A) the same as one 22% inflationary increase
B) the same as one 30% inflationary increase
C) the same as an inflationary increase of 10% followed by an inflationary increase of 20%
D) less than an inflationary increase of 10% followed by an inflationary increase of 20%
E) more than an inflationary increase of 10% followed by an inflationary increase of 20%

Chapter 2

Wonderful Math Recipes

Four out of every three persons have trouble with fractions.

—Anonymous

Overlap Scenarios

☞ Tip #14: **Situations involving either A or B do not necessarily preclude the possibility of both A and B. When two events overlap, the number of items in both groups will exceed 100% due to mutual inclusivity.**

Snapshot

Group problems (also called overlap problems or Venn-diagram problems) illustrate situations in which there is crossover among sets of information. These types of problems may be thought of as membership problems—some members do this, some members do that, some members do both, and some members do neither.

Consider the following exchange:

Travel Agent: Of people who booked through our agency and traveled to Australia or New Zealand this past winter, 65% went to Australia and 55% went to New Zealand.

Customer: That doesn't sound right. You should check your numbers. How can more than 100% of your tourists travel to Australia or New Zealand?

The customer's response is likely invalid, because it is entirely possible that some tourists traveled to both Australia and New Zealand during their winter vacation.

How can we work with this kind of information in order to get a clear picture of the number of people in each category? The key lies in manipulating the following formula:

Two-Groups Formula		Solve
Add:	Members in Group A	x
Add:	Members in Group B	x
Less:	Both	<x>
Add:	Neither	x
	Total	xx

In Las Vegas, 21 tourists were surveyed, and it was found that 15 of these people liked to play the slot machines and 7 people liked to play table games such as blackjack. Only 4 people surveyed said they played neither slot machines nor table games. How many people surveyed played both slot machines and table games?

Two-Groups Formula		Solve
Add:	Slot Machines	15
Add:	Table Games	7
Less:	Both	<?>
Add:	Neither	4
	Total	21

Answer: Five people played both slot machines and table games.

Here is the visual representation of the Las Vegas scenario.

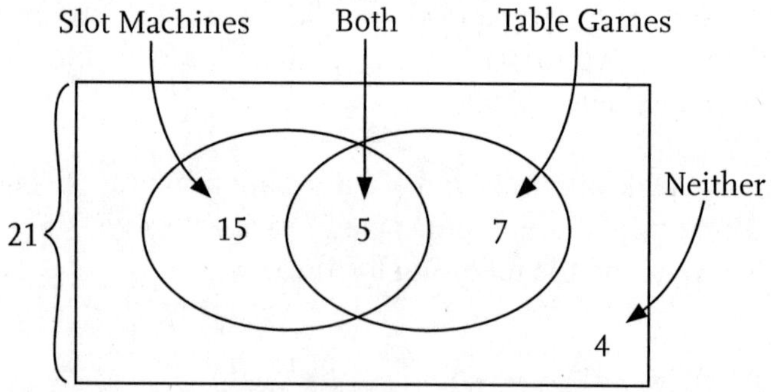

Calculation:

Group A + Group B – Both + Neither = Total
$15 + 7 - x + 4 = 21$
$26 - x = 21$
$x = 5$

The rectangular region represents 100 percent—the whole pie, so to speak. The total of all the numbers inside the rectangle cannot add up to more than 21 people. This is why "neither" must be added back; it falls outside the circle but within the rectangle. "Both" must be subtracted out because it was counted twice, once when counting members of Group A and again when counting members of Group B.

Summary of Formula Templates

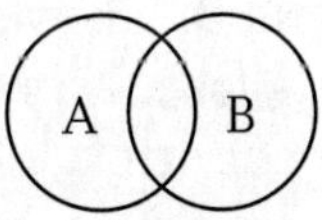

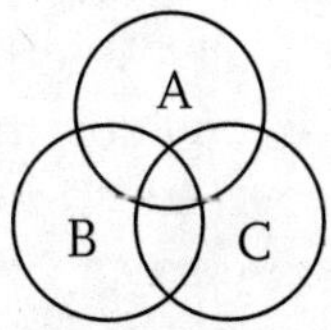

Two-Groups Formula		Solve
Add:	Group A	x
Add:	Group B	x
Less:	Both	<x>
Add:	Neither	x
	Total	xx

Three-Groups Formula		Solve
Add:	Group A	x
Add:	Group B	x
Add:	Group C	x
Less:	A & B	<x>
Less:	A & C	<x>
Less:	B & C	<x>
Add:	All of A & B & C	x
Add:	None of A or B or C	x
	Total	xx

NOTE ✍ Let's briefly contrast what is commonly referred to as the "inclusive or" versus the "exclusive or." Consider the statement "Today's weather will be rainy or windy." In casual conversation this may be interpreted to mean that the weather is anticipated to be either rainy or windy, but not both rainy and windy. In strict logic, however, such situations involving either A or B do not necessarily preclude the possibility of both A and B. In other words, the weather could indeed be both rainy and windy. There is an implied "both" in this particular type of "either-or" statement. The multiple-choice problems encountered in this section also involve the use of the "inclusive or." That is, these problems provide for the possibility of both A and B.

Other situations may involve the use of the "exclusive or." Here it is not possible to have a situation involving both A and B. Take for example the student who must choose to enroll in either high school A or high school B. Obviously the student cannot enroll at both schools. We effectively understand that the choice is between A or B, and does not include both A and B. Likewise, the negative statement "no eating or drinking allowed" must be logically interpreted as "we must neither eat nor drink." We cannot interpret this sign to mean that because we can't eat or drink, we can therefore eat <u>and</u> drink simultaneously in order to avoid the "no eating <u>or</u> drinking" mandate.

In summary, situations involving the use of the "exclusive or" lead to *mutually exclusive* options. Situations involving the use of the "inclusive or" lead to *mutually inclusive* options. Note also that we may refer to two options that are mutually inclusive as two options that are not mutually exclusive.

45. Country Club ()

A country club has 2,500 members, 700 of whom play tennis and 500 of whom play golf. If 1,500 members play neither tennis nor golf, how many members play both tennis and golf?

A) 200
B) 1,000
C) 1,300
D) 2,200
E) 2,800

46. Nova vs. Rebound ()

Jill: In our survey, 53% of those questioned said they used Nova running shoes for jogging and 47% said they used Rebound running shoes for jogging.

Jack: In that case everyone questioned used one product or the other.

Jill: No, 24% said they didn't use either one.

Jill's statements imply that

A) Not all of those questioned liked to go jogging.

B) Some of those who said they used either Nova or Rebound running shoes for jogging were lying.

C) Some of those who used neither product were nevertheless familiar with Nova and Rebound running shoes.

D) Some of those who used Nova running shoes for jogging were among those who also used Rebound running shoes for jogging.

E) Many Nova and Rebound running shoes are actually made in the same factories in various international locations.

47. Science ( )

Each of 35 students is enrolled in chemistry, physics, or both. If 22 are enrolled in chemistry and 19 are enrolled in physics, how many are enrolled in both chemistry and physics?

A) 3
B) 6
C) 9
D) 12
E) 29

48. Standardized Test ( )

If 85% of the test takers taking an old paper and pencil standardized exam answered the first question on a given math section correctly, and 75% of the test takers answered the second question correctly, and 65% of the test takers answered both questions correctly, what percent answered neither correctly?

A) 5%
B) 10%
C) 15%
D) 20%
E) 25%

49. Language Classes ()

According to the admissions and records office of a major university, the schedules of X first-year college students were inspected and it was found that S number of students were taking a Spanish course, F number of students were taking a French course, and B number of students were taking both a Spanish and a French course. Which of the following expressions gives the percentage of students whose schedules were inspected who were taking neither a Spanish course nor a French course?

A) $100 \times \dfrac{X}{B + F + S}$

B) $100 \times \dfrac{B + F + S}{X}$

C) $100 \times \dfrac{X - F - S}{X}$

D) $100 \times \dfrac{X + B - F - S}{X}$

E) $100 \times \dfrac{X - B - F - S}{X}$

50. Valley High ()

At Valley High School, four students have chosen to do an independent study project in biology, four in chemistry, and four in physics. Two students have chosen to do independent study projects in both biology and chemistry, two in both chemistry and physics, and two in both biology and physics. One student has chosen to do an independent study project in all three of biology, chemistry, and physics. How many students are involved in the above-mentioned independent study projects?

A) 5
B) 7
C) 9
D) 12
E) 19

51. German Cars ( )

The New Marketing Journal conducted a survey of wealthy German car owners. According to the survey, all wealthy car owners surveyed owned one or more of the following three brands: BMW, Mercedes, or Porsche. Respondents' answers were grouped as follows: 45 owned BMW cars, 38 owned Mercedes cars, and 27 owned Porsche cars. Of these, 15 owned both BMW and Mercedes cars, 12 owned both Mercedes and Porsche cars, 8 owned both BMW and Porsche cars, and 5 persons owned all three types of cars. How many different individuals were surveyed?

A) 70
B) 75
C) 80
D) 110
E) 130

Matrix Scenarios

Tip #15: A matrix is a marvelous tool for summarizing data that is being contrasted across two variables and that can be sorted into four distinct outcomes.

Snapshot

A matrix is a box-like table. In fact, a simple matrix is a table with exactly nine boxes—four boxes on the inside and five boxes on the outside. Our job with respect to matrix problems is to fill in known information and, through simple arithmetic, find the unknown information. A completed matrix will show data in all nine boxes.

Take, for example, a batch of 100 toys, fresh off the production line. Each toy has exactly two of four characteristics: It is either blue or green and either large or small. The total number of toys equals 100, and this figure appears in the bottom right corner of the extended matrix. The boxes with dotted lines provide useful extensions for use in totaling data.

Factory Toys

In our batch of 100 toys, 55 are blue and the rest are green. Half the toys are large and half the toys are small. If 15 of the toys are small and green, how many are large and blue?

Here is a matrix, shown as a nine-box table, which is used to set up this problem:

Color

Size		**Blue**	**Green**	
	Large	x	x	xx
	Small	x	x	xx
		xx	xx	xxx

Let's fill in the given numbers:

Color

		Blue	Green	
Size	Large	?		50
	Small		15	50
		55	45	100

Finally, we total down and across:

Color

		Blue	Green	
Size	Large	?=20	30	50
	Small	35	15	50
		55	45	100

Why do matrixes work so neatly? Things work neatly as long as all data is mutually exclusive and collectively exhaustive. What does this mean? Mutually exclusive is a fancy way of saying that the data does not overlap; it is distinct. In other words, toys must be either blue or green and either large or small. We can't have toys which are both blue and green (for example, blue-green or striped) or both large and small (that is, medium-sized). Collectively exhaustive means that the number of data is finite.

There are only 100 toys, of which exactly 55 are blue, 45 are green, 50 are large, and 50 are small. The fact that this data is mutually exclusive and collectively exhaustive allows everything to total both down and across.

Because matrixes handle information so neatly, it is not surprising that they are a consultant's favorite presentation tool. Legend has it that one young consultant became so enamored with matrixes that he called them "boxes of joy"!

52. Job Search ( )

Of 35 applicants applying for a job, 20 had at least 7 years of work experience, 23 had degrees, and 3 had less than 7 years of work experience and did not have a degree. How many of the applicants had at least 7 years of work experience and a degree?

A) 3
B) 9
C) 11
D) 12
E) 20

53. Nordic ( )

In a certain town in Norway, 75% of the people have blond hair, 65% have blue eyes, and 45% have both blond hair and blue eyes. What percentage of the people in the town have neither blond hair nor blue eyes?

A) 5%
B) 20%
C) 30%
D) 45%
E) 65%

54. Singles ( )

In a graduate physics course, 70% of the students are male and 30% of the students are married. If two-sevenths of the male students are married, what fraction of the female students are single?

A) $\frac{2}{7}$ B) $\frac{1}{3}$ C) $\frac{1}{2}$ D) $\frac{2}{3}$ E) $\frac{5}{7}$

55. Batteries ( )

One-fifth of the batteries produced by a certain upstart factory are defective, and one-quarter of all batteries produced are rejected by the quality control technician. If one-tenth of the non-defective batteries are rejected by mistake, and if all the batteries not rejected are sold, then what percentage of the batteries sold by the factory are defective?

A) 4%
B) 5%
C) 6%
D) 8%
E) 12%

56. Experiment ( )

Sixty percent of the rats included in an experiment were female. If some of the rats died during an experiment and 70% of the rats that died were male rats, what was the ratio of the death rate among the male rats to the death rate among the female rats?

A) 7:2
B) 7:3
C) 2:7
D) 3:7
E) Cannot be determined from the information given

Mixture Scenarios

Tip #16: **Mixture problems are best solved using the barrel method, which enables information to be summarized in a 3-row by 3-column "table."**

Snapshot

There are essentially two kinds of mixture problems: dry mixtures and wet mixtures. Dry mixtures involve dry ingredients, such as candies, nuts, or coffee beans, whereas wet mixtures involve liquids such as chemicals. The amount or quantity of a mixture is usually measured in liters, gallons, pounds, or kilograms. Below is a template, herein called the barrel method, for use in solving wet or dry mixture problems.

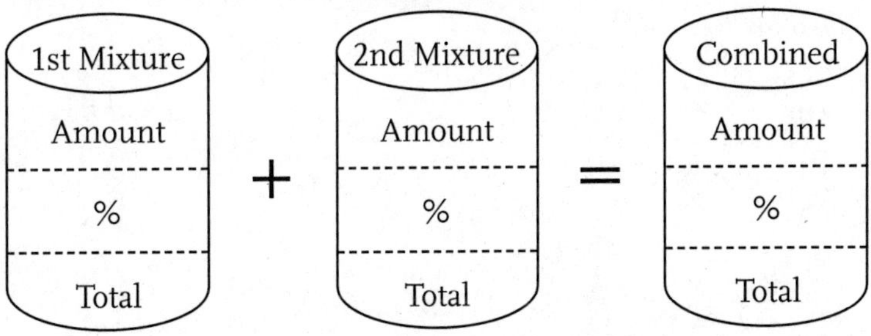

As illustrated in the next example, this technique works miraculously in helping to structure mixture problems. Example: How many liters of pure acid must be added to 40 liters of a 10% solution of acid to make a 25% acid solution?

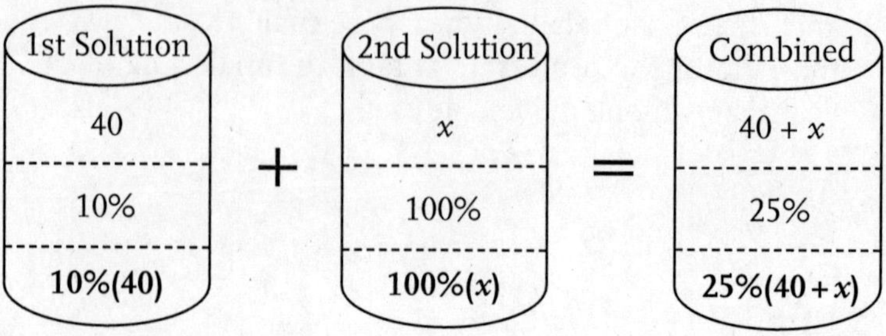

Next, we create an equation from the totals (in bold) appearing at the bottom of the barrels:

$$10\%(40) + 100\%(x) = 25\%(40 + x)$$

$$4 + x = 10 + 0.25x$$

$$1.0x - 0.25x = 10 - 4$$

$$0.75x = 6$$

$$x = \frac{6}{0.75}$$

$$x = 8 \text{ liters}$$

57. Petroleum ( )

50 gallons of a 40% oil mixture is added to 25 gallons of a 20% oil mixture. What percentage of the resulting mixture is oil?

A) 25%
B) 30%
C) $33\frac{1}{3}\%$
D) 35%
E) 45%

58. Perfect Blend ( )

A dealer wishes to mix 20 kilograms of coffee selling for \$7.50 per kilogram with some more expensive coffee selling for \$15.00 per kilogram to make a mixture that will sell to coffee houses for \$12.00 per kilogram. How many kilograms of the more expensive coffee should be used?

A) 25
B) 30
C) 35
D) 40
E) 45

59. Nuts ( )

A wholesaler sells 100 pounds of mixed nuts at \$2.50 a pound. She mixes peanuts worth \$1.50 a pound with cashews worth \$4.00 a pound. How many pounds of cashews does she use?

A) 40
B) 45
C) 50
D) 55
E) 60

60. Adding H_2O ( )

How many gallons of water must be added to 40 gallons of a solution that is 30% salt to dilute it to a solution that is 25% salt?

A) 6
B) 8
C) 9
D) 10
E) 12

61. Evaporation ( )

How many liters of water must be evaporated from 50 liters of a 3% sugar solution to get a 10% sugar solution?

A) 15
B) $16\frac{2}{3}$
C) 27
D) $33\frac{1}{3}$
E) 35

62. Gold ( )

An alloy weighing 24 ounces is 70% gold. How many ounces of pure gold must be added to create an alloy that is 90% gold?

A) 6
B) 9
C) 12
D) 24
E) 48

Weighted Average Scenarios

☞ Tip #17: **To find a weighted average, multiply each event by its associated weight and total the results. In the case of probabilities, multiply each event by its respective probability and total the results.**

Snapshot

The weighted average concept is actually quite intuitive. To find weighted average, we multiply "events" by their respective "weight" and total the results. Events are the things that we wish to rate, rank, or judge. Weights refer to the amount of emphasis we want to attribute to each event and are commonly expressed as percentages, fractions, decimals, or probabilities. The beauty of weighted average is that we can assign different weights based on the relative importance of events—the more important the event, the more weight it is given.

Below is the weighted average formula for two events:

$$\text{WA} = (\text{Event}_1 \times \text{Weight}_1) + (\text{Event}_2 \times \text{Weight}_2)$$

An alternative format, for ease of computation, unfolds as follows:

$$\begin{aligned} \text{Event}_1 \times \text{Weight}_1 &= xx \\ \text{Event}_2 \times \text{Weight}_2 &= \underline{xx} \\ & \underline{\underline{xx}} \end{aligned}$$

Exam Time

A student scores 60 out of 100 points on his midterm exam and 90 out of 100 points on his final exam. If the exams are both weighted equally, counting for 50% of the student's final course grade, then what is his course grade?

$$60 \times 50\% = 30$$
$$90 \times 50\% = \underline{45}$$
$$\underline{\underline{75}}$$

Based on the same information above, what is the student's final course grade if the midterm exam is weighted 40 percent and the final exam is weighted 60 percent?

$$60 \times 40\% = 24$$
$$90 \times 60\% = \underline{54}$$
$$\underline{\underline{78}}$$

Note that the weights above could also have been expressed in terms of fractions or decimals:

$$60 \times \frac{40}{100} = 24 \qquad\qquad 60 \times 0.4 = 24$$
$$90 \times \frac{60}{100} = \underline{54} \qquad\qquad 90 \times 0.6 = \underline{54}$$
$$\underline{\underline{78}} \qquad\qquad \underline{\underline{78}}$$

Hiring and promotion decisions are classic examples of situations in which subjective influences can override an objective decision-making process. Weighted ranking therefore presents a method to quantify decision opportunities.

Consider a company with 10 salespersons, one of whom is to be named National Sales Manager. The 10 candidates are first ranked from 1 to 10 (10 being the highest rating) across three criteria. The three criteria—technical skills, people skills, and track record—are weighted using the weights of 0.2, 0.3 and 0.5, respectively. Note that instead of using the decimals 0.2, 0.3, and 0.5, we could also use the percentages 20%, 30%, 50%, the fractions $\frac{2}{10}$, $\frac{3}{10}$, $\frac{5}{10}$, or even whole numbers such as 2, 3, and 5.

Based on the results from weighting the data, Sabrina receives the highest ranking, while George gets the next highest ranking. The weights used will typically total to 1, as is so often the case when dealing with percentages, fractions, decimals, or probabilities. Sometimes problems will use arbitrary weights which do not equal 1.

Chess

In chess, a pawn is worth 1 point, a knight or bishop is worth 3 points, a rook is worth 5 points, and a queen is worth 9 points. Player A has captured two rooks, a knight, and 3 pawns. Player B has captured a bishop, 4 pawns, and a queen. Who is ahead and by how much?

	Player A	Player B
Pawns:	$3 \times 1 = 3$ pts	$4 \times 1 = 4$ pts
Bishops:		$1 \times 3 = 3$ pts
Knights:	$1 \times 3 = 3$ pts	
Rooks:	$2 \times 5 = 10$ pts	
Queens:		$1 \times 9 = 9$ pts
	16 points	16 points

Answer: Both players are tied at 16 points each.

EXHIBIT 2.1 PERFORMANCE OF SALESPERSONS

	Technical skills and product knowledge	People skills and ability to communicate	Track record and ability to get things done
Albert	2	3	7
Betty	5	2	6
George	6	5	9
Jed	3	7	1
Jono	8	10	3
Martha	10	1	2
Patricia	1	4	8
Randy	9	9	4
Sabrina	4	6	10
William	7	8	5

EXHIBIT 2.2 PERFORMANCE USING WEIGHTED AVERAGE

	Technical skills and product knowledge Weight = 0.2	People skills and ability to communicate Weight = 0.3	Track record and ability to get things done Weight = 0.5	Total
Albert	2 × 0.2 = 0.4	3 × 0.3 = 0.9	7 × 0.5 = 3.5	4.8
Betty	5 × 0.2 = 1.0	2 × 0.3 = 0.6	6 × 0.5 = 3.0	4.6
George	6 × 0.2 = 1.2	5 × 0.3 = 1.5	9 × 0.5 = 4.5	7.2
Jed	3 × 0.2 = 0.6	7 × 0.3 = 2.1	1 × 0.5 = 0.5	3.2
Jono	8 × 0.2 = 1.6	10 × 0.3 = 3.0	3 × 0.5 = 1.5	6.1
Martha	10 × 0.2 = 2.0	1 × 0.3 = 0.3	2 × 0.5 = 1.0	3.3
Patricia	1 × 0.2 = 0.2	4 × 0.3 = 1.2	8 × 0.5 = 4.0	5.4
Randy	9 × 0.2 = 1.8	9 × 0.3 = 2.7	4 × 0.5 = 2.0	6.5
Sabrina	4 × 0.2 = 0.8	6 × 0.3 = 1.8	10 × 0.5 = 5.0	7.6
William	7 × 0.2 = 1.4	8 × 0.3 = 2.4	5 × 0.5 = 2.5	6.3

EXHIBIT 2.3 RANKING OF SALESPERSONS

	Technical skills and product knowledge Weight = 0.2	People skills and ability to communicate Weight = 0.3	Track record and ability to get things done Weight = 0.5	Total score (The higher the better)	Rank (The lower the better)
Sabrina	4 × 0.2 = 0.8	6 × 0.3 = 1.8	10 × 0.5 = 5.0	7.6	1
George	6 × 0.2 = 1.2	5 × 0.3 = 1.5	9 × 0.5 = 4.5	7.2	2
Randy	9 × 0.2 = 1.8	9 × 0.3 = 2.7	4 × 0.5 = 2.0	6.5	3
William	7 × 0.2 = 1.4	8 × 0.3 = 2.4	5 × 0.5 = 2.5	6.3	4
Jono	8 × 0.2 = 1.6	10 × 0.3 = 3.0	3 × 0.5 = 1.5	6.1	5
Patricia	1 × 0.2 = 0.2	4 × 0.3 = 1.2	8 × 0.5 = 4.0	5.4	6
Albert	2 × 0.2 = 0.4	3 × 0.3 = 0.9	7 × 0.5 = 3.5	4.8	7
Betty	5 × 0.2 = 1.0	2 × 0.3 = 0.6	6 × 0.5 = 3.0	4.6	8
Martha	10 × 0.2 = 2.0	1 × 0.3 = 0.3	2 × 0.5 = 1.0	3.3	9
Jed	3 × 0.2 = 0.6	7 × 0.3 = 2.1	1 × 0.5 = 0.5	3.2	10

Sweet Sixteen

On her 16th birthday, Jane received \$500 from each of her two uncles. Both amounts were deposited in two local banks, one bank paying 6 percent per annum and the other paying 7 percent per annum. How much in total did she earn from these two investments over the course of exactly one year?

$$\$500 \times 6\% = \$30$$
$$\$500 \times 7\% = \underline{\$35}$$
$$\underline{\underline{\$65}}$$

63. Bowling ( )

In a bowling tournament, 60% of the bowlers had an average (arithmetic mean) score of 110, and the other 40% had an average score of 130. If each of the bowlers bowled an equal number of games, what is the average (arithmetic mean) score of all the bowlers in the tournament?

A) 115
B) 118
C) 120
D) 122
E) 125

64. The Right Mix ( )

Chocolate A costs $9 per kilogram and chocolate B costs $15 per kilogram. What is the average cost per kilogram of a combined mixture that is 80% chocolate A and 20% chocolate B?

A) $9.80
B) $10.00
C) $10.20
D) $12.00
E) $13.80

65. Baseball ( )

A baseball bat and a catcher's mitt currently have the same price. If the price of a baseball bat rises 3% and the price of a catcher's mitt rises 5%, how much more (in percentage terms) will it cost to buy 3 baseball bats and 3 catcher's mitts?

A) 4%
B) 8%
C) 16%
D) 24%
E) 48%

66. Venture Capital ( )

The probability that a certain new venture will generate a return of $5,000,000 is estimated to be 20%. The probability that the new venture will generate a return of $1,500,000 is 50%. The probability that the new venture will generate a return of $500,000 is 30%. Based on a weighted-average assessment, how much money can be expected to be earned?

A) $1,850,000
B) $1,900,000
C) $1,950,000
D) $2,000,000
E) $2,050,000

67. Portfolio ( )

An investor is looking at three different investment possibilities. The first investment opportunity has a $\frac{1}{6}$ chance of returning $90,000, a $\frac{1}{2}$ chance of returning $50,000, and a $\frac{1}{3}$ chance of losing $60,000. A second investment opportunity has a $\frac{1}{2}$ chance of returning $100,000 and a $\frac{1}{2}$ chance of losing $50,000. The third investment opportunity has a $\frac{1}{4}$ chance of returning $100,000, a $\frac{1}{4}$ chance of returning $60,000, a $\frac{1}{4}$ chance of losing $40,000, and a $\frac{1}{4}$ chance of losing $80,000. Assuming the investor chooses to invest in all three investments, what will be his or her expected return?

A) $55,000
B) $60,000
C) $65,000
D) $70,000
E) $75,000

Chapter 3

Favorite Numeracy Dishes

There are three kinds of people in this world: those who can count and those who can't.

—Anonymous

Markup vs. Margin

☞ Tip #18: Markup is always larger than margin.

Snapshot

The following story highlights the important real-life distinction between markup and margin:

A boss told his sales clerk to put price tags on all newly arrived toys and transfer them to the sales floor. The boss said, in mumbled language, that he wanted to adhere to company policy that requires earning a 25% return on all goods sold. However, the clerk heard "25%" and, looking at the cost of each toy item, promptly marked the goods up 25% based on the cost sheets. The company accountant eventually discovered this mistake when investigating why overall profit had fallen. The confusion created was the result of failing to distinguish between markup and margin. Markup is always based on product cost and gross margin is based on selling price. In short, markup is always a larger number than margin. Mathematically, markup is based on product cost—a smaller number than sales price—and consequently must be a larger percentage. Margin is based on selling price—a larger number—and consequently results in a smaller percentage. In other words, the markup required to earn a 25% gross margin is $33\frac{1}{3}$%, not 25%. The clerk should have multiplied the cost of each good by 1.33, not 1.25.

Margin and markup are commonly expressed as percentages. Thus, margin is calculated as the difference between sales price and product cost divided by sales price, which is then multiplied by 100. Markup is calculated as the difference between sales price and product cost divided by product cost, which is then multiplied by 100.

Four formulas are used when dealing with margin and markup. The first two are used when working with dollars (or other currencies) in order to calculate margin or markup percentages:

$$\text{Margin \%} = \frac{\text{Sales (\$)} - \text{Cost (\$)}}{\text{Sales (\$)}} \times 100\%$$

$$\text{Markup \%} = \frac{\text{Sales (\$)} - \text{Cost (\$)}}{\text{Cost (\$)}} \times 100\%$$

The next two formulas are used when working directly with percentages. Here's the formula for converting markup percentage to gross margin percentage:

$$\text{Margin \%} = \frac{\text{Markup \%}}{100\% + \text{Markup \%}}$$

The following formula is used to convert gross margin percentage to markup percentage. Note that the terms "margin" and "gross margin" may be used interchangeably.

$$\text{Markup \%} = \frac{\text{GM \%}}{100\% - \text{GM \%}}$$

Multipliers

If we want to achieve a margin of 16.67%, then we mark up goods by 20%. Or put another way, if goods are marked up by 20%, then a margin of 16.67% will be achieved. Of course, the easy way to arrive at this percentage is to use the "multipliers" given in *Exhibit 3.1.* To achieve a margin of 16.67%, we simply multiply product cost by 1.20.

$$\text{Markup \%} = \frac{\text{GM \%}}{100\% - \text{GM \%}}$$

$$\text{Markup \%} = \frac{16.67\%}{100\% - 16.67\%} = \frac{16.67\%}{83.33\%} = 20\%$$

$$\text{Margin \%} = \frac{\text{Markup \%}}{100\% + \text{Markup \%}}$$

$$\text{Margin \%} = \frac{20\%}{100\% + 20\%} = \frac{20\%}{120\%} = 16.67\%$$

EXHIBIT 3.1 COMMON MARGIN AND MARKUP PERCENTAGES

Margin	Markup	Multipliers – Multiply product cost by these factors to achieve target margin percentages
$16\frac{2}{3}\%$	20%	1.20
20%	25%	1.25
25%	$33\frac{1}{3}\%$	1.33
$33\frac{1}{3}\%$	50%	1.50
50%	100%	2.00

68. Gross Margin ( )

If a company buys a product for $7.50 and sells the product for $10, what is its gross margin percentage?

A) 20%
B) 25%
C) $33\frac{1}{3}\%$
D) 50%
E) 100%

69. Markup ( )

If a company buys a product for \$7.50 and sells the product for \$10, what is its markup percentage?

A) 20%
B) 25%
C) $33\frac{1}{3}\%$
D) 50%
E) 100%

70. Alpha Inc. ( )

Alpha Inc. habitually marks its products up 50%. How much is the corporation making on each product sold with respect to a gross margin percentage?

A) 20%
B) 25%
C) $33\frac{1}{3}\%$
D) 50%
E) 100%

71. Beta Corp. ( )

Beta Corp. wants to sell its products and achieve a gross margin of 20%. What should the markup percentage be to achieve this gross margin percentage?

A) 20%
B) 25%
C) $33\frac{1}{3}\%$
D) 50%
E) 100%

Price, Cost, Volume, and Profit

☞ Tip #19: **An increase (decrease) in sales volume does not necessarily translate to an increase (decrease) in profit.**

Snapshot

Price-cost-volume-profit scenarios highlight the interplay of the three profit components: price, cost, and volume. Just because a computer store is selling more computers than last year does not mean that its profits are up. Why? Because an increase (decrease) in sales volume (that is, number of unit sales) of a good or service does not necessarily equal an increase (decrease) in revenues or profits. Likewise, an increase (decrease) in the price of a good or service does not necessarily equal an increase (decrease) in profits. Moreover, an increase (or decrease) in the cost of a good or service does not necessarily translate to a decrease (or increase) in profits.

Note that the words "expense" and "cost" are deemed synonymous; the words "volume" and "number of units" are interchangeable.

Here are three pertinent formulas:

Cost Formula

Cost per unit × Number of units = Cost

$7 per unit × 1,000 units = $7,000

Revenue Formula

Price per unit × Number of units = Revenue

$10 per unit × 1,000 units = $10,000

Profit Formula

Revenue – Cost = Profit

\$10,000 – \$7,000 = \$3,000

Two variations of profit formulas include:

(Price per unit – Cost per unit) × No. of units = Profit

(\$10 per unit – \$7 per unit) × 1,000 units = \$3,000

(Price per unit × No. of units) – (Cost per unit × No. of units) = Profit

(\$10 per unit × 1,000 units) – (\$7 per unit × 1,000 units) = \$3,000

72. Century Bus Line ( )

The Century Bus Line's financial forecast for the coming year indicates a decrease in overall operating profits. However, the company forecasts a record number of paying passengers for the coming year. Which of the following facts, A through E below, would best contribute to an explanation of this scenario?

A) The average length of a Century Bus Line trip will decrease while the average number of trips will increase.

B) There is to be no projected increase in the number of canceled trips in the coming year.

C) Two new competitors of Century Bus Line are expected to enter the market in the coming year.

D) Century Bus Line will cut ticket prices by an average of 25% in the coming year.

E) Century Bus Line will be able to reduce significantly its overall operating costs due to government deregulation of fuel in the coming year.

73. High-Speed Boat

To operate a high-speed boat, it costs g cents a mile for gasoline and h cents a mile for oil. How many dollars will it cost, in terms of gasoline and oil, to run the high-speed boat for 100 miles?

A) g
B) $g + h$
C) $g + 0.1h$
D) $100g + 100h$
E) $(g + h)/100$

74. Cases

A manufacturer of cassette cases wants to make a profit of x dollars. When it sells 10,000 cassette cases it costs 5 cents (=$0.05) to make the first 1,000 cases, and then it costs y cents a case to make the remaining 9,000 cassette cases. What price in dollars should it charge for the 10,000 cassette cases?

A) $50 + 10y + x$
B) $50 + 90y + x$
C) $10{,}000 + 1{,}000y$
D) $10{,}000 + 9.000y$
E) $10{,}000 + 1{,}000y + 100x$

75. Delicatessen ( )

A large delicatessen purchased p pounds of cheese for c dollars per pound. If d pounds of the cheese had to be discarded due to spoilage and the delicatessen sold the rest for s dollars per pound, which of the following represents the gross profit on the sale of the purchase (gross profit equals sales revenue minus product cost)?

A) $(s - c)(p - d)$
B) $s(p - d) - cp$
C) $c(p - d) - sd$
D) $d(s - c) - pc$
E) $pc - ds$

☞ **Tip #20: When price-cost-volume-profit problems are presented in the form of algebraic expressions, it is best to first find a dollar per unit, dollar per person, usage per unit, or usage per person figure.**

Snapshot

The statement, "c pounds of carrots cost 20 dollars" results in a dollar per unit figure expressed as 20 dollars/c pounds or 20/c dollars per pound. In these types of problems, note how the dollar figure becomes the fraction's numerator, not its denominator. Using the cost formula, we solve the problem below as follows:

Cost = Cost per unit × Number of units

$$\text{Cost} = \frac{20 \text{ dollars}}{c \cancel{\text{ pounds}}} \times N \cancel{\text{ pounds}}$$

$$\text{Cost} = \frac{20N}{c} \text{ dollars}$$

76. Paper Clips ( )

If p paper clips cost c cents, how many cents do q paper clips cost?

A) $\frac{cq}{p}$ B) $\frac{pq}{c}$ C) $\frac{p}{cq}$ D) $\frac{c}{pq}$ E) pcq

77. Garments ( )

If s shirts can be purchased for d dollars, how many shirts can be purchased for t dollars?

A) sdt B) $\frac{ts}{d}$ C) $\frac{td}{s}$ D) $\frac{d}{st}$ E) $\frac{s}{dt}$

78. Copy ( )

If each photocopy of a manuscript costs 2 cents per page, what is the cost, in cents, to reproduce y copies of a y-page manuscript?

A) $2y$
B) y^2
C) $2y^2$
D) $4y$
E) $4y^2$

79. Pete's Pet Shop (🌶🌶)

At Pete's Pet Shop, 35 cups of bird seed are used every 7 days to feed 15 parakeets. How many cups of bird seed would be required to feed 9 parakeets for 12 days?

A) 32
B) 36
C) 39
D) 42
E) 45

Break-Even Point

☞ Tip #21: **In break-even point problems, break-even occurs exactly where variable revenue (sales revenue less variable costs) equals total fixed costs.**

Snapshot

Break-even point is defined as the point where zero profit is achieved. That is, the point where sales revenue exactly equals total costs. Total cost is divided into fixed and variable components because this facilitates break-even point calculations.

Variable costs vary directly with units produced, whereas fixed costs do not. Fixed costs therefore stay the same regardless of whether you sell one item or a million items.

In everyday living, rent is considered a fixed cost, whereas utilities (water, gas, and electricity usage) are typically considered variable costs. In terms of rent, it does not matter how much you use your apartment in one month—the cost is the same. In terms of utilities, the more you use in one month, the more you pay.

Suppose that company A produces widget Z. Assume that a company produces only widget Zs and has fixed costs of $4,500, a selling price of $10 per unit, and a variable cost of $2.50 per unit. What is its break-even point in *units*?

$$\text{BE}_{\text{units}} = \frac{\text{Fixed Costs}}{\text{Selling Price}_{\text{per unit}} - \text{Variable Cost}_{\text{per unit}}}$$

$$\text{Break Even}_{\text{units}} = \frac{\$4{,}500}{\$10 - \$2.50}$$

$$\text{Break Even}_{\text{units}} = \frac{\$4{,}500}{\$7.50} = 600 \text{ units}$$

What if we wanted instead to determine the break-even point in terms of sales *dollars*?

$$\text{Break Even}_{\text{dollars}} = \frac{\text{Fixed Costs}}{\dfrac{\text{SP}_{\text{per unit}} - \text{VC}_{\text{per unit}}}{\text{SP}_{\text{per unit}}}}$$

$$\text{Break Even}_{\text{dollars}} = \frac{\$4{,}500}{\dfrac{\$10 - \$2.50}{\$10}}$$

$$\text{Break Even}_{\text{dollars}} = \frac{\$4{,}500}{\dfrac{\$7.50}{\$10}}$$

$$\text{Break Even}_{\text{dollars}} = \frac{\$4{,}500}{0.75} = \$6{,}000$$

Of course, we can prove all of this by remembering that break-even is that point where total revenues equal total costs.

Total Revenue = Total Costs

Total Revenue = Fixed Costs + Variable Costs

600 units × $10 per unit = $4,500 + (600 units × $2.50 per unit)

$6,000 = $4,500 + $1,500

$6,000 = $6,000

Break-even point theory is sometimes applied to situations involving competing projects. Different projects invariably have different cost and revenue structures. When two projects are being evaluated on competing terms, we may want to know how many additional sales are needed before the costs, profits, or revenues of one project become equal to the other.

To solve for situations involving competing projects, first set the two projects "equal" to one another. Next, fill in given information, and solve for the remaining variable.

$$\text{Revenue}_{\text{Project 1}} = \text{Revenue}_{\text{Project 2}}$$

Price per unit × No. of units = Price per unit × No. of units

$$\text{Cost}_{\text{Project 1}} = \text{Cost}_{\text{Project 2}}$$

Cost per unit × No. of units = Cost per unit × No. of units

$$\text{Profit}_{\text{Project 1}} = \text{Profit}_{\text{Project 2}}$$

Profit per unit × No. of units = Profit per unit × No. of units

80. Book Publisher ( )

The chart below lists the price and cost data for publication of a soon-to-be released paperback novel. What is the minimum number of copies that must be sold in order for the publisher not to lose money on the book?

Bookstore Price	$5.95/copy
Fixed Costs	
Author fees (advance)	$150,000
Editing and design	$20,000
Advertising and promotion	$230,000
Variable Costs	
Paper, binding and printing	33¢/copy
Storage and distribution	12¢/copy
Author's royalties	50¢/copy

A) 20,000
B) 48,358
C) 50,000
D) 67,143
E) 80,000

81. Sabrina to Change Jobs ( )

Sabrina is contemplating a job switch. She is thinking of leaving her job paying \$85,000 per year to accept a sales job paying \$45,000 per year plus 15% commission for each sale made. If each of her sales is for \$1,500, what is the smallest number of sales she must make per year if she is not to lose money because of the job change?

A) 26
B) 177
C) 178
D) 200
E) 600

Aggregate Costs vs. Per-Unit Costs

☞ Tip #22: **Total costs must be distinguished from per-unit costs, as is the case whenever the number of units varies between two scenarios under comparison.**

Snapshot

A distinction must be drawn between aggregate (or total costs) and per-unit costs because total costs often are compared without regard to the number of units involved. We have to be careful not to assume that the total number of units is constant. For example, a single piece of Dove chocolates and a single piece of Valentine chocolates are of equivalent size and of comparable quality. Since a box of Dove chocolates cost £5 (British pounds) and a box of Valentine chocolates costs £4 (British pounds), can we conclude Valentine chocolates are our best buy? Not necessarily—what if a box of Dove chocolates contains significantly more chocolates compared with a box of Valentine chocolates?

When making per-unit comparisons between two scenarios, it is important to visualize the three components of the cost formula:

Scenario 1	*Scenario 2*
$\frac{\text{Total Costs}}{\text{Units}} = \text{Per unit cost}$	$\frac{\text{Total Costs}}{\text{Units}} = \text{Per unit cost}$

82. Grapes ( )

It costs much less to grow an acre of grapes in California than to grow an acre of watermelons in Oklahoma. This fact should be obvious to anyone who read last year's annual farm report, which clearly shows that many millions of dollars more were spent last year growing watermelons in Oklahoma than were spent growing grapes in California.

Which of the following, if true, would most seriously call into question the reasoning above?

A) Profits on the sale of California grapes are much higher than the profits derived from the sale of Oklahoma watermelons.

B) Far fewer total acres of grapes were grown in California last year than total acres of watermelons in Oklahoma.

C) An acre of grapes in Oklahoma costs much more to grow than an acre of watermelons in California.

D) Part of the California-grown grapes were used to feed livestock, whereas all of the watermelons grown in Oklahoma were used for human consumption.

E) State subsidies accounted for a larger percentage of the amount spent growing watermelons in Oklahoma than of the amount spent growing grapes in California last year.

Efficiency

☞ **Tip #23: Efficiency is output divided by input. Cost efficiency is cost divided by efficiency.**

Snapshot

Although efficiency can be defined in a few different ways, perhaps the most multi-purpose definition appears below. Input is usually measured in units.

$$\text{Efficiency} = \frac{\text{Output}}{\text{Input}}$$

$$\text{Cost Efficiency} = \frac{\text{Cost}}{\text{Efficiency}} = \frac{\text{Cost}}{\frac{\text{Output}}{\text{Input}}}$$

Suppose that there are two employees working at a garment company. Employee A can make 10 garments in 5 hours. Employee B can make 5 garments in 5 hours. Their efficiencies become:

Employee A

$$\text{Efficiency} = \frac{\text{Output}}{\text{Input}} = \frac{10 \text{ garments}}{5 \text{ hours}} = 2 \text{ garments per hour}$$

Employee B

$$\text{Efficiency} = \frac{\text{Output}}{\text{Input}} = \frac{5 \text{ garments}}{5 \text{ hours}} = 1 \text{ garment per hour}$$

We can conclude that employee A is twice as efficient as employee B. But is employee A more cost efficient than employee B? Let's say that employee A is paid $40 per hour but employee B is paid $10 per hour.

Employee A

$$\text{Cost Efficiency} = \frac{\$40/\text{hour}}{\frac{10 \text{ garments}}{5 \text{ hours}}} = \$20 \text{ per garment}$$

Employee B

$$\text{Cost Efficiency} = \frac{\$10/\text{hour}}{\frac{5 \text{ garments}}{5 \text{ hours}}} = \$10 \text{ per garment}$$

So we can conclude that although employee A is twice as efficient as employee B, employee B is twice as cost-effective as employee A. Of course, we did make some assumptions. For example, we assumed that the quality of the craftsmanship is comparable between the two employees and that the slower employee B did not incur any additional costs when taking more hours to finish his or her work (that is, factory overhead).

83. Act-Fast ( )

One Act-Fast tablet contains twice the pain reliever found in a tablet of regular aspirin. A consumer will have to take two aspirin to get the relief provided by one Act-Fast tablet. And since a bottle of Act-Fast costs the same as a bottle of regular aspirin, consumers can be expected to switch to Act-Fast.

Which of the following, if true, would most weaken the argument that consumers will be discontinuing the use of regular aspirin and switching to Act-Fast?

A) A regular aspirin tablet is less than one-half as large as an Act-Fast tablet.

B) Neither regular aspirin nor Act-Fast is as effective in relief of serious pain as are drugs available only by prescription.

C) Some headache sufferers experience a brief period of nausea shortly after taking Act-Fast but not after taking regular aspirin.

D) A bottle of regular aspirin contains more than twice as many tablets as does a bottle of Act-Fast.

E) The pain reliever in Act-Fast is essentially the same pain reliever found in regular aspirin.

84. Prototype (🌶🌶🌶)

A prototype fuel-efficient car (P-Car) is estimated to get 80% more miles per gallon of gasoline than does a traditional fuel-efficient car (T-Car). However, the P-Car requires a special type of gasoline that costs 20% more per gallon than does the gasoline used by a T-Car. If the two cars are driven the same distance, what percent less than the money spent on gasoline for the T-Car is the money spent on gasoline for the P-Car?

A) $16\frac{2}{3}\%$

B) $33\frac{1}{3}\%$

C) 50%

D) 60%

E) $66\frac{2}{3}\%$

Distribution and Allocation

☞ Tip #24: **Elements within distributions may be uniformly spaced or concentrated in "bunches." Allocations may be equal or unequal, proportionate or disproportionate.**

Snapshot

If we have 75 cans of pop to be distributed among 15 students, we will likely assume that each student will receive 5 cans of pop. If a house has 3 rooms covering an aggregate area of 750 square feet, it is unlikely that we would assume that each room is exactly 250 square feet. If a bookstore sells 5 dictionaries in 5 days, we may or may not be tempted to assume that a single book was sold each day. Perhaps multiple books were sold on one or more days while no books were sold on other days. In short, distributions are not always proportional or linear.

85. Wanda ( )

Wanda is working on a take-home math test which contains 4 sections and a total of 100 true-false problems. If each problem is worth 1 point, and Wanda has finished the first 3 sections as well as the first problem of the fourth section, has she finished more than 75% of the problems on her test?

A) Yes
B) No
C) Can't Tell

86. Noble Book Club ( )

A sales representative from the Noble Book Club recently hailed the club's free CD-ROM gift program as a big boost for sales. A year ago, the club began offering a free CD-ROM gift for any member ordering four or more books in a single month. Since then, the number of members ordering four or more books has risen by nearly 40%.

Which one of the following, if true, would most seriously weaken the representative's assessment of the program's effect on sales?

A) The number of members ordering fewer than four books in a single month also rose in the last year.

B) Most members ordered the same number of books over the year but concentrated their orders in specific months.

C) The cost of providing free CD-ROM gifts nearly offset the increased revenue from higher sales.

D) Most competing book clubs, many of which also sponsored free gift programs, saw a drop in sales last year.

E) The membership in book clubs across the country rose by more than 15% over the past year.

☞ Tip #25: **Front-loading occurs when a greater proportion of inputs (for example, time, materials, costs, efforts, benefits) are entered early in a process. Back-loading occurs when a greater proportion of inputs are entered late in a process.**

Snapshot

Examples of front-loading and back-loading pervade real-life settings. If we use all of our vacation time early in the work year, we have front-loaded our vacation time and have no vacation days left. If we save all our vacation time for the end of the year, our decision is to back-load our vacation days.

Another example stems from scholastic study habits. Each of us can recall, with some humor, a time when we crammed insanely for a final exam. If we plotted our study time over the course of an entire semester (or quarter) for a given course, a huge spike would indicate a disproportional amount of study time recorded late in the school term. This exemplifies back-loaded study, where the inputs of time and effort occur late in the process.

Now reflect on the traditional formal education process in developed countries. A person completes his or her formal schooling in a very concentrated manner, regardless of whether it ends with high school, community college, or university. There is typically little formal schooling after that. The assumption is that the kinds of skills that are learned in school (or are supposed to be learned) are to benefit a person throughout his or her entire life—certainly basic skills such as reading, writing, arithmetic, and reasoning. After all, formal study ends somewhere between the ages of 18 and 25, and yet the average life expectancy of a person living in developed countries is around 80 years of age. Formal education may be viewed as a front-loaded endeavor insofar as the inputs of time and effort occur early in the process.

87. Max Motorcycles ( )

An advertisement designed to convince readers of the great durability of motorcycles manufactured by the Max Motorcycle Company cites as evidence the fact that half of all motorcycles built by the company in the last 10 years are still on the road today compared to no more than a third for any other manufacturer of motorcycles.

Which of the following, if true, most strongly supports the advertisement's argument?

A) After taking inflation into account, a new Max Motorcycle costs only slightly more than a new model did 10 years ago.

B) In the past 10 years, Max Motorcycles has made fewer changes in the motorcycles it manufactures compared with changes made by other motorcycle companies.

C) Owners of Max Motorcycles typically keep their motorcycles well maintained.

D) The number of motorcycles built by Max Motorcycle each year has not increased sharply in the past 10 years.

E) Max Motorcycles have been selling at relatively stable prices in recent years.

Chapter 4

Special Math Garnishments

Math is like love. It starts out simple, then it gets complicated.

—Anonymous

Translating Basic Graphs

☞ Tip #26: **Forward-slashing graphs indicate that two variables are directly related. Back-slashing graphs indicate that two variables are indirectly (inversely) related. Graphs formed with straight lines have variables which are linearly related. Graphs formed with curved lines have variables that are non-linearly related.**

Snapshot

Each of the twelve line graphs included in *Exhibit 4.1* has the following characteristics: It is either *flat* (horizontal) or *sloping* (upward or downward) and it is either composed of *straight lines* or *curves*.

All the graphs in *Exhibit 4.1* showcase two variables—money and time. Money appears on the vertical or y-axis, and time appears on the horizontal or x-axis. Graphs that are *flat* represent graphs with variables that do not vary with one another. On the other hand, graphs that are *sloping* represent variables that do vary with one another. If a graph *slopes upward* (whether a straight line or a curve), the two variables are said to be directly related to one another. This means that as one variable increases the other also increases or, conversely, as one variable decreases, the other deceases. Graphs that *slope downward* represent variables that are indirectly related to one another. This means that as one variable increases, the other decreases or, conversely, as one variable decreases the other increases.

Finally, graphs that consist of *straight lines* have variables that are linearly related. "Linear" means that the variables increase or decrease proportionally. Graphs that consist of *curved lines* have variables that are non-linearly related. "Non-linear" means that the variables increase or decrease disproportionately.

Graphs composed of *straight lines sloping upward* depict variables that are both directly and linearly related to one another. For every unit of increase in one variable, there is a proportional unit of increase in the other variable. A good real-life example might involve the painting of a room. For every hour you spend painting, you get one more unit of work done.

Graphs composed of *curved lines sloping upward* depict variables that are directly but non-linearly related to one another. For every unit of increase in one variable, there is more than one unit of increase in the other variable. A good real-life example might involve completing a difficult puzzle with many pieces. For each additional hour we spend working on it, we complete a disproportionate amount of the puzzle because the pieces start to fit.

Graphs composed of *straight lines sloping downward* depict variables that are indirectly but linearly related to one another. For every unit of increase in one variable, there is a proportional unit of decrease in the other variable. For every hour you spend painting your room, you get one less hour of leisure time.

Graphs composed of *curved lines sloping downward* depict variables that are both indirectly and non-linearly related to one another. For every unit of increase in one variable, there is a disproportionately larger decrease in the other variable. A good real-life example might involve study time and retention over a sustained study period. For every hour spent studying, we gain a unit of benefit. But after two or three hours, we experience diminishing returns for our time and effort spent.

88. Quiz on Basic Graphs ( )

Use the following twelve graphs in *Exhibit 4.1* and match each graph (A through L) with one of the twelve statements that follow on pages 112–113. For your convenience, graphs are divided into three categories: rising graphs, flat graphs, and falling graphs.

Exhibit 4.1 Basic Graphs

Rising Graphs (A,B,C,D) $ = wages/salary

Flat Graphs (E,F,G,H) $ = wage/salary

Falling Graphs (I,J,K,L) $ = wage/salary

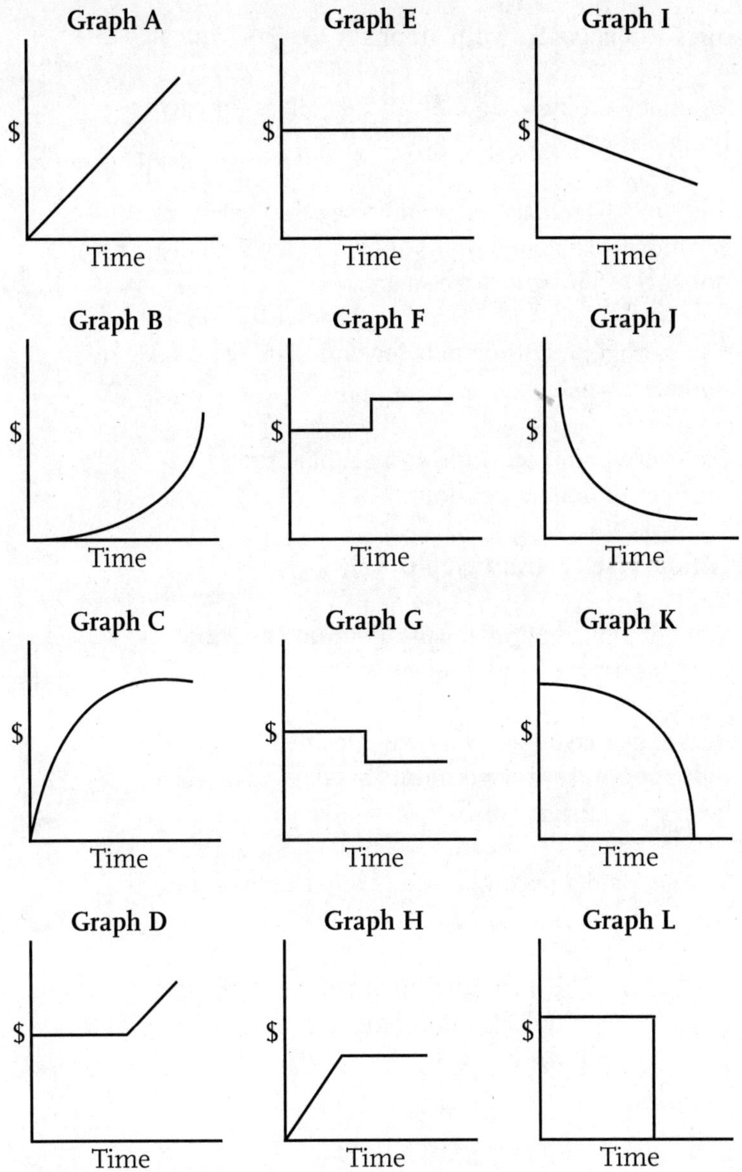

The topic is one that is dear to us all—salaries and wages. A one-chili rating is assigned to this problem as you only need to match six out of twelve statements correctly in order to get the whole problem right!

Rising Graphs (select from graphs A, B, C or D):

You get paid incrementally more dollars for each hour you work. ❑

You get a fixed base salary but may also get additional pay based on hours worked which are not covered by your base salary. ❑

You get paid incrementally fewer dollars for each hour you work. ❑

For every hour you work, you get paid the same number of dollars per hour. ❑

Flat Graphs (select from graphs E, F, G or H):

You got a promotion at work that increased the fixed salary you now receive. ❑

Because of layoffs at work, you decided to take a reduction in your fixed monthly salary in order to keep your current job. ❑

You get paid a fixed (flat) wage regardless of the hours you work. ❑

You get paid by the hour, but cannot earn more than a certain dollar amount during the course of your working week. ❑

Falling Graphs (select from graphs I, J, K or L):

Your yearly pension decreases in terms of real dollars during each year of your retirement. ❑

Your wages decrease gradually with age until the time you stop working altogether. ❑

Your wages are fixed but stop the moment you retire. ❑

Your wages decrease dramatically with age, but you can still earn a little in your retirement. ❑

Line Graphs, Pie Charts, and Bar Charts

☞ Tip #27: **Line graphs are best used to show how data changes over time. Pie charts are great for breaking data down by category. Bar charts are ideal for side-by-side comparisons of up to six pieces of data. Knowing the strength of each type of graph or chart and how they are best used to present information also helps us in interpreting the underlying data.**

Snapshot

Three basic types of graphs and charts are commonly used to present data. The following questions would precipitate the use of each of these graphs or charts in the commercial realm:

Line graphs

"How have corporate profits, sales, or earnings per share increased or decreased over time?" (Hint: A line graph is especially useful for showing trends.)

Pie charts

"How are corporate expenses broken down by major categories including the percentage of total corporate expenses that are administrative, salaries and wages, advertising, operating, and miscellaneous?" (Hint: Pie charts are especially useful for breaking things down.)

Bar charts

"How do the different corporate divisions stack up with one another in terms of head count (total number of employees)?" or "How does our company rank alongside its major competitors in terms of market share?" (Hint: Bar charts are most useful for showing side-by-side comparisons.)

Here is how these three types of visual tools might be used in a non-work setting:

Line graphs

"Am I watching too much television?" (Hint: After charting a week-by-week record of hours of TV viewing at home, you use the data to create a line graph showing the number of TV-free hours at home per week.)

Pie charts

"Where is all my money going?" (Hint: Use a pie chart to display what percentage of your disposable income is spent on different categories.)

Bar charts

"What are my greatest personal strengths and how do they rank?" (Hint: After identifying a half-dozen of your most salient personal or professional strengths, you assign values to them from 1 to 10—1 being lowest and 10 being the highest—and display them using a bar chart.)

Use the following two graphs to answer questions 89–91 regarding Grand Hotel & Casino's revenues for the month of January.

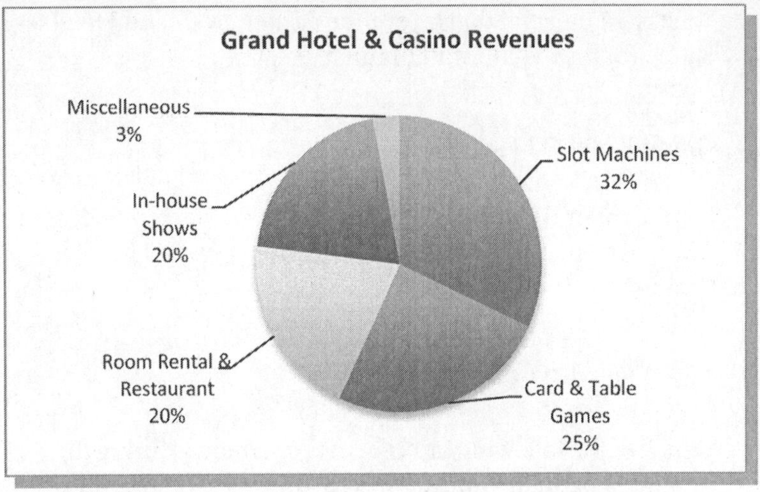

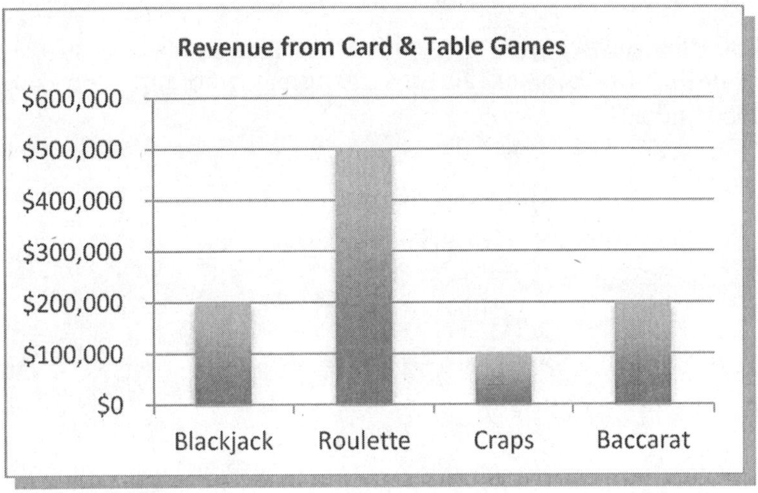

89. Grand Hotel Q1

What was the total dollar revenue earned by Grand Hotel & Casino for the month of January?

90. Grand Hotel Q2

For the month of January, revenue from *blackjack* provided what percent of the total revenue for Grand Hotel & Casino?

91. Grand Hotel Q3

Which of the following brought in more money during the month of January—*miscellaneous* revenues or revenues from *craps*?

Use the following two graphs to answer questions 92–94 regarding the composition of Argentina's workforce between 1980 and 2010.

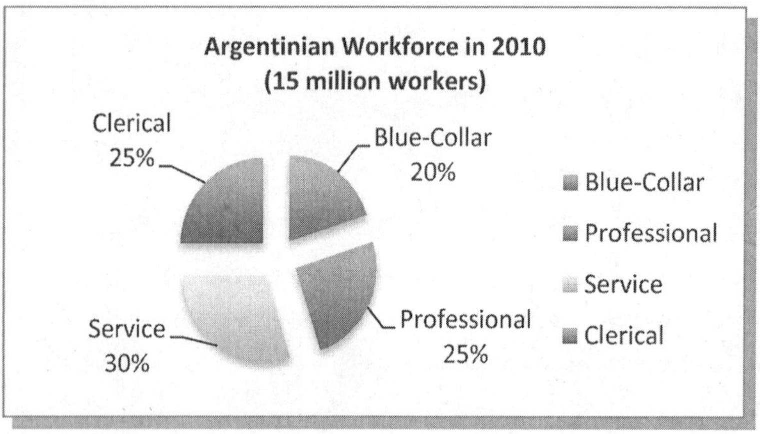

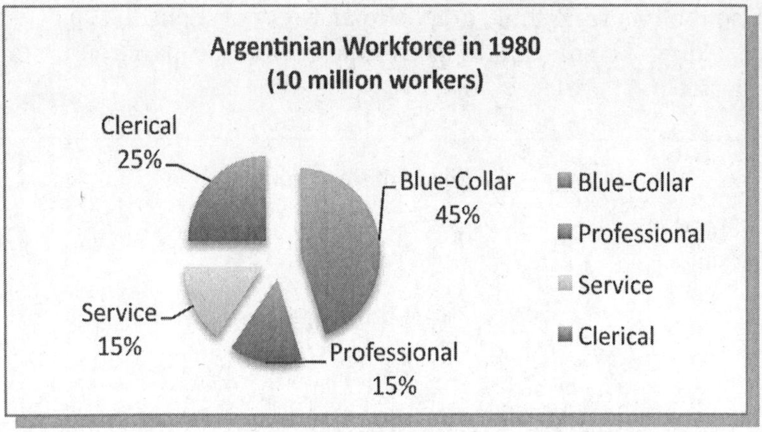

92. Workforce Q1 ()

The decrease in blue-collar workers from 45% in 1980 to 20% in 2010 represents a decrease of how many millions of people?

93. Workforce Q2 ()

According to the data presented, the same number of people worked in the service sector in 2010 as worked in which sector in 1980?

94. Workforce Q3 (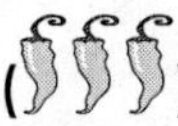)

From 1980 to 2010, the number of Argentinians working in the service, professional, and clerical sectors increased. Calculate the percentage increases for each of these sectors in terms of the number of workers. Also, calculate the percentage decrease for the blue-collar sector in terms of the number of workers. Which sectors, if any, experienced a percentage increase that was greater than the percentage decrease attributed to the blue-collar sector?

Use the following three graphs to answer questions 95–96 regarding Magna Fund's performance for the past four quarters.

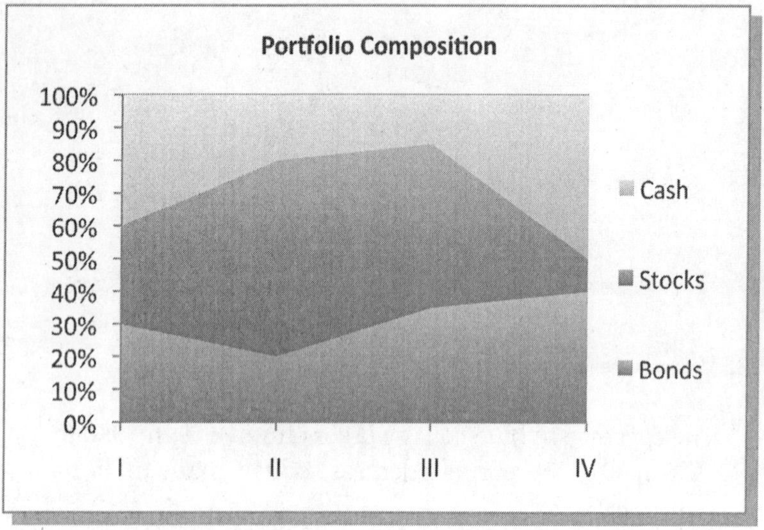

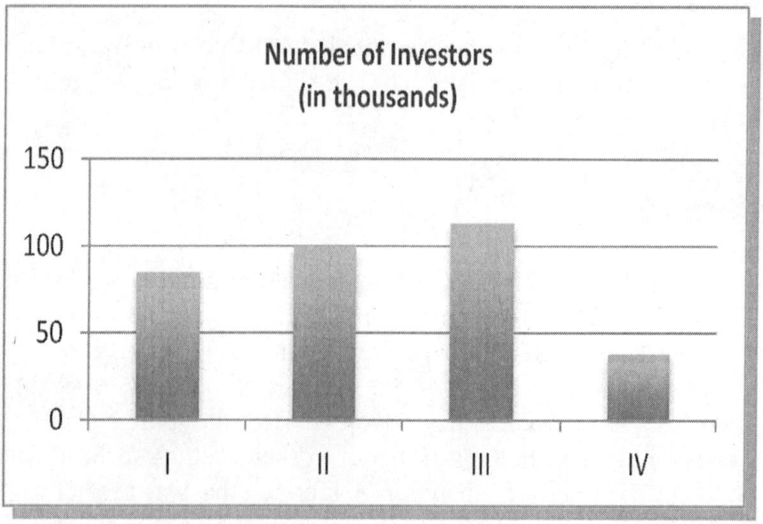

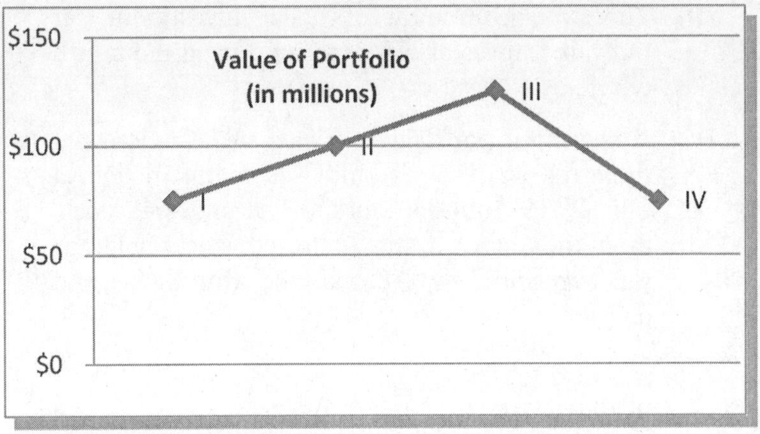

95. Magna Fund Q1 ( )

What was the percentage increase in the portfolio's value from the first quarter to the third quarter?

A) 40%

B) 50%

C) $66\frac{2}{3}\%$

D) 125%

E) $166\frac{2}{3}\%$

96. Magna Fund Q2 ( )

Which of the following statements can be inferred from the three graphs above?

I. The percent of the portfolio accounted for by cash declined from the third to the fourth quarter.

II. Bonds accounted for one-third or more of the portfolio in three of four quarters.

III. The average amount of dollar investment per individual investor was greatest during the fourth quarter.

IV. If the stock portfolio consisted of common and preferred shares only, and was maintained in a ratio of 70% common stock to 30% preferred stock, then the value of the preferred stock holdings was approximately $18 million during the second quarter.

A) I & II
B) II & III
C) III & IV
D) II, III & IV
E) All of the above

Correlation

☞ Tip #28: **Positive correlations have graphs sloping upward; negative correlations have graphs sloping downward. The stronger the correlation between two variables, the nearer to a straight line will be their graph; the graph of a perfect correlation is a straight line.**

Snapshot

Correlation is about the relationship between two (or more) variables. By relationship we are referring not only to how they are connected but also to the strength of their connection. Two variables can be connected in two basic ways: positively and negatively. Positively correlated variables are directly related. Negatively correlated variables are indirectly related.

The strength of the connection between two variables is measured by a value ranging from +1 to –1. A correlation of +1 is a positive-perfect correlation; a correlation of –1 is a negative-perfect correlation. In real life, positive-perfect or negative-perfect correlations are rare. Note that although the scatter points of perfect correlations appear as straight lines, positive-high correlations look like fat fountain pens, and positive-low correlations are shaped like squashes. *Exhibit 4.2* provides a flowchart to summarize the statistical relationships as they relate to correlation. *Exhibit 4.3* illustrates the different types of correlation using scatter plot graphs, which plot the intersection of data points.

EXHIBIT 4.2 CORRELATION FLOWCHART

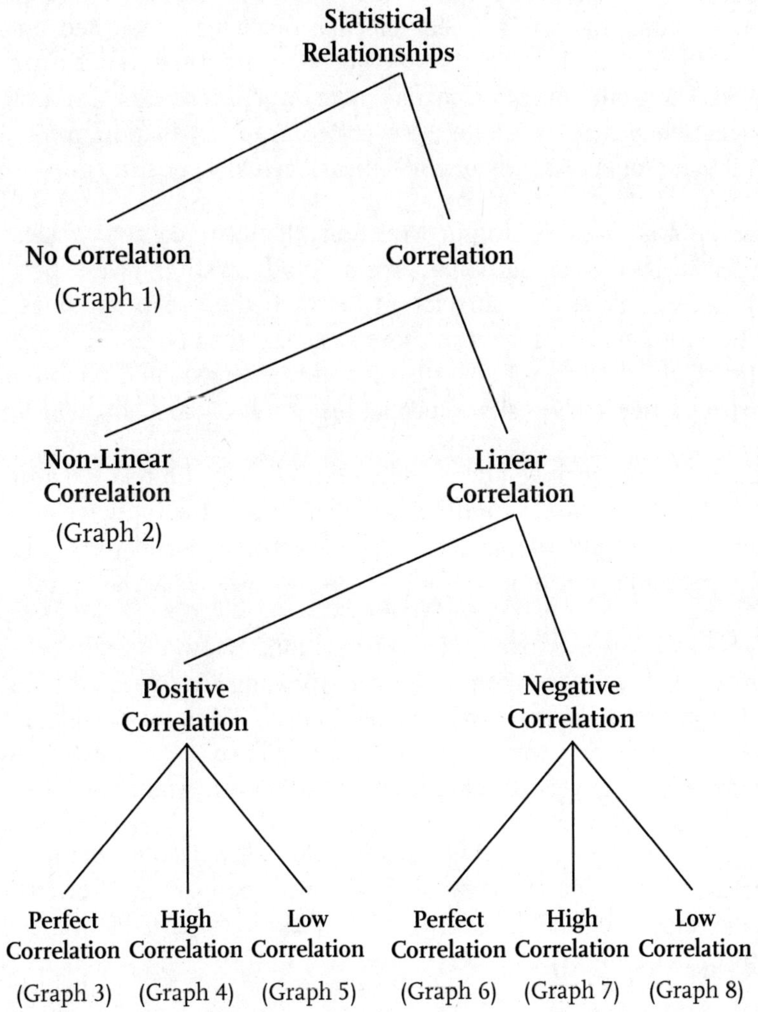

Graph 1 indicates no correlation. "What is the relationship between hair color and longevity?" We would expect the scatter points describing such data to be completely random.

Graph 2 depicts a non-linear correlation. "How is performance affected by anxiety levels?" Whether it be an academic or athletic undertaking, a moderate level of anxiety is optimal, with low or high levels of anxiety suboptimal.

Graph 3 is a positive-perfect correlation. The amount of coffee beans used and the number of cups of coffee consumed are directly related. Note that it would not be accurate to substitute the phrase "amount of coffee beans grown or produced" for "amount of coffee beans used" because coffee crop production might fluctuate independently of coffee beans actually consumed.

Graph 4 is a positive-high correlation. "How are company sales affected by increasing dollars spent on advertising?" We expect to see a very strong correlation between these two variables. The more we advertise, the more sales increase (within certain limits). In the case of advertising and sales, we would typically expect to see a 0.8 positive correlation.

Graph 5 is a positive-low correlation. "What is the relationship between height and weight?" The taller we are, the more we will likely weigh, but the correlation between height and weight is not particularly strong.

Graph 6 is a negative-perfect correlation. "How does the life experience we gain with each year's passing vary with respect to the number of years we have left to live?" Well, for every year we live and gain experience, we have one fewer year to live. It's a trade-off: one year of experience traded for one year of our life.

Graph 7 is a negative-high correlation. "How is our memory after drinking lots of beer?" It is a pretty good bet that for any individual, the more he or she drinks, the less remarkable will be his or her memory.

Graph 8 is a negative-low correlation. "How is one's level of physical fitness linked to food consumption?" There might be some truth to the idea that the more we eat the less fit we are. But we also know of cases where highly fit individuals eat copious amounts of food, as is the case with many professional athletes, particularly those playing contact sports such as American football or rugby.

EXHIBIT 4.3 BASIC CORRELATION GRAPHS

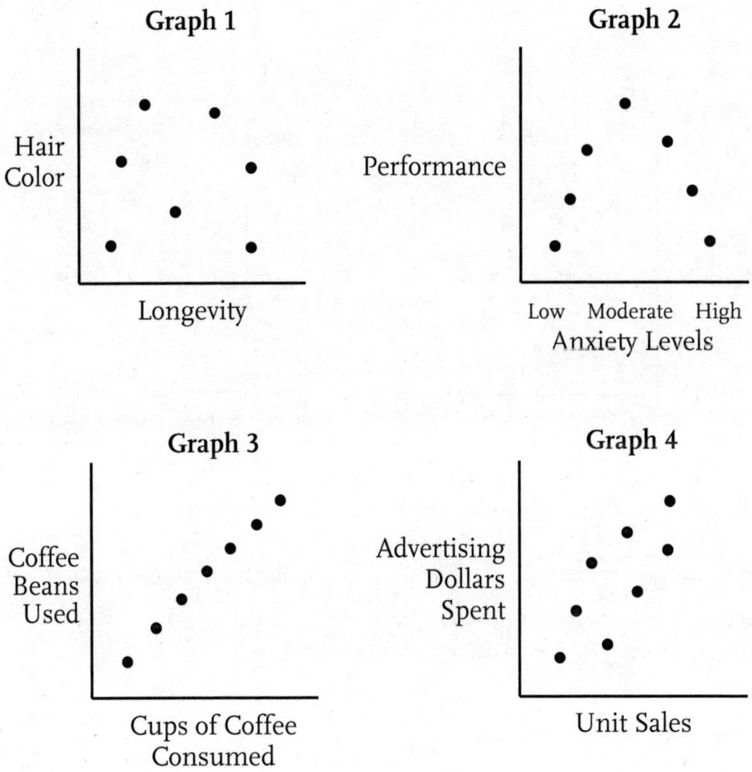

EXHIBIT 4.3 (Cont'd)

Graph 5

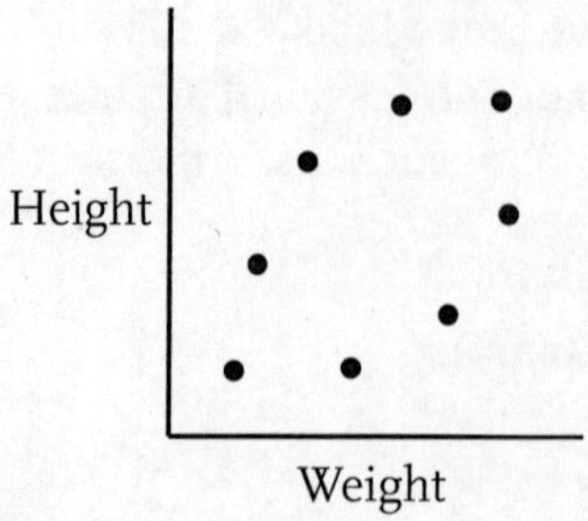

Graph 6

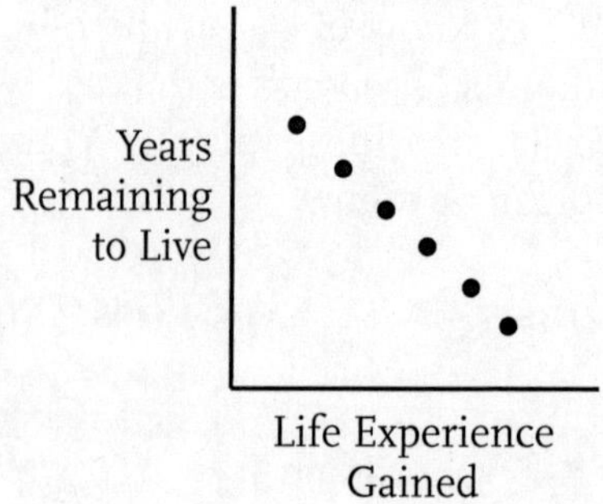

Graph 7

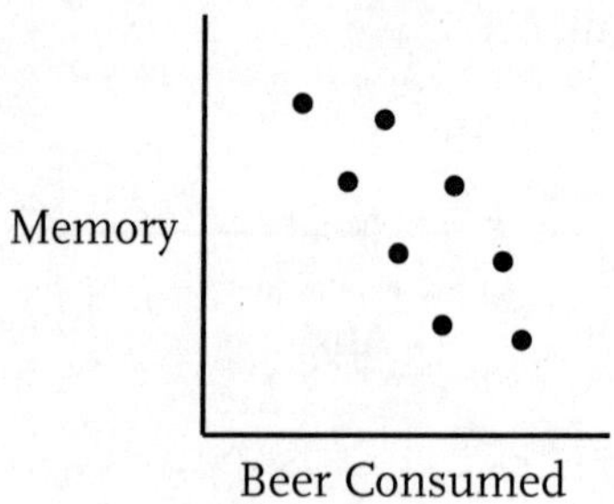

Graph 8

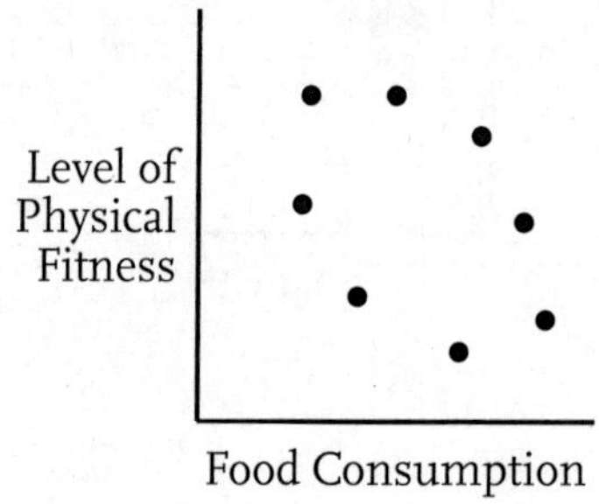

97. Quiz on Correlation Analysis ( )

Placing the appropriate letter in each box, match graphs A through H, on the following page, with the caption below that best describes the type of correlation depicted.

Positive-perfect correlation	☐	Negative-perfect correlation	☐
Positive-high correlation	☐	Negative-high correlation	☐
Positive-low correlation	☐	Negative-low correlation	☐
Non-linear correlation	☐	Zero correlation	☐

Graph A

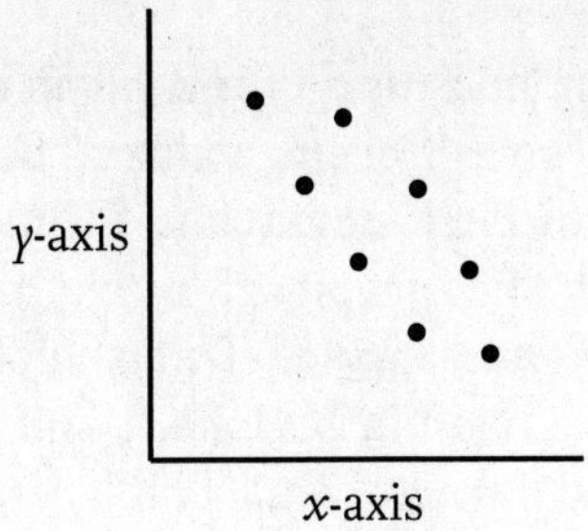

Graph B

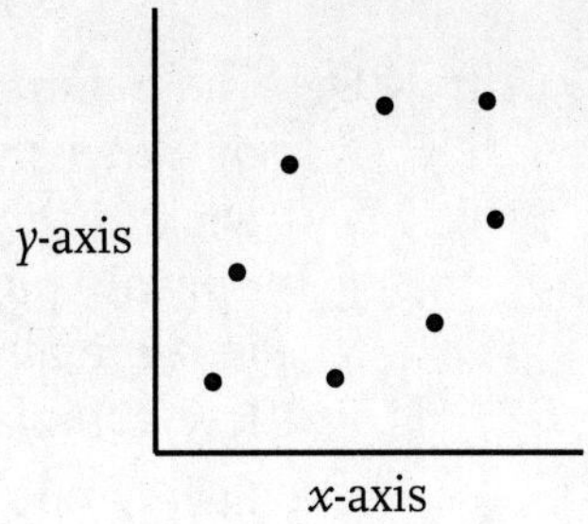

Graph C

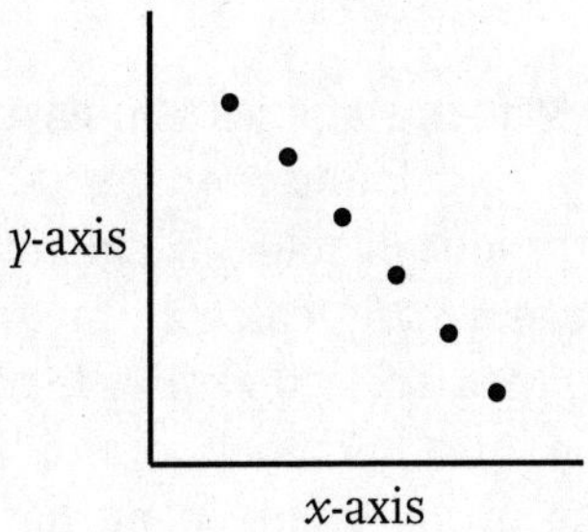

Graph D

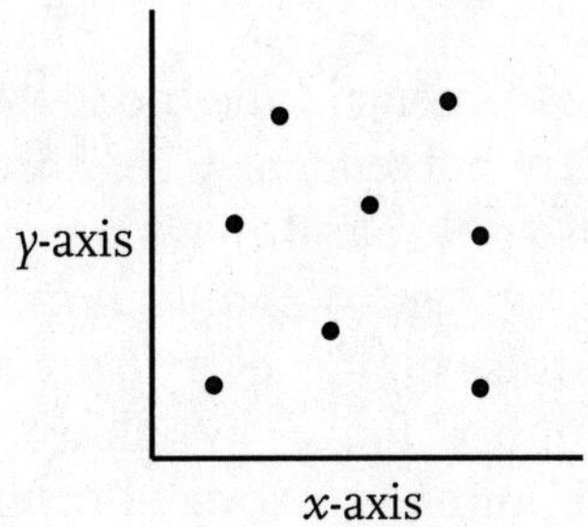

Graph E

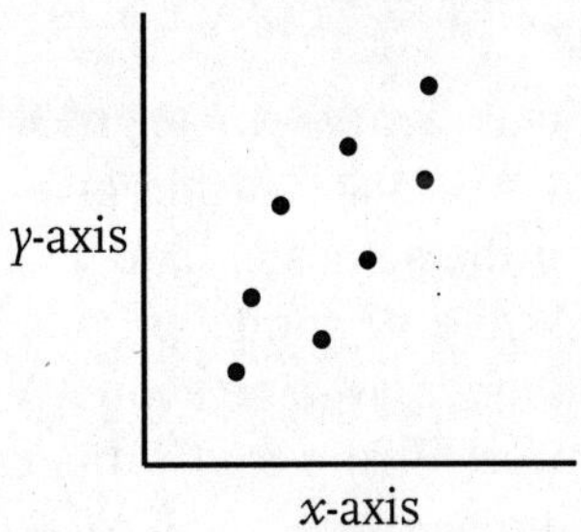

Graph F

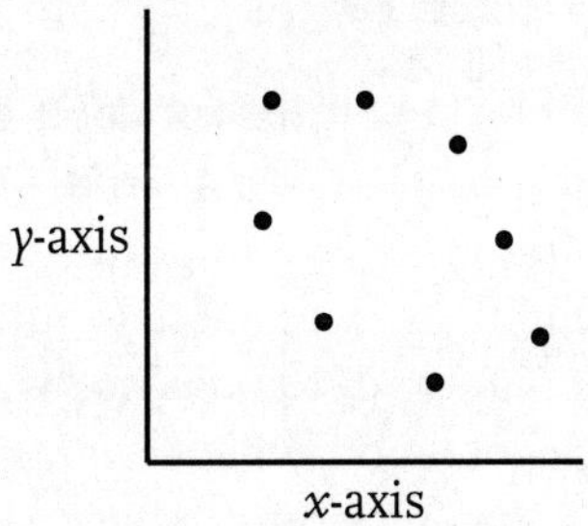

Graph G

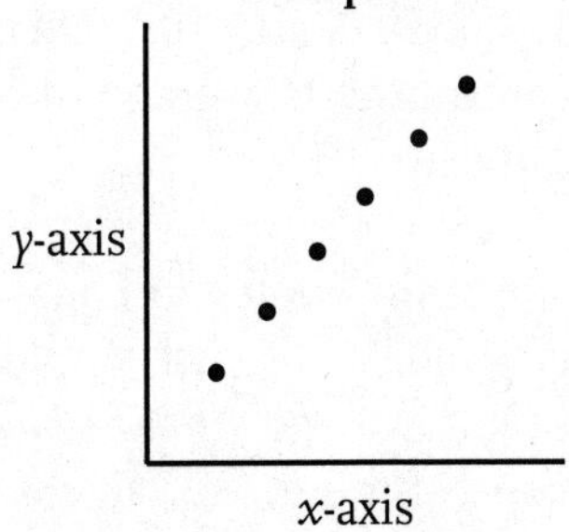

Graph H

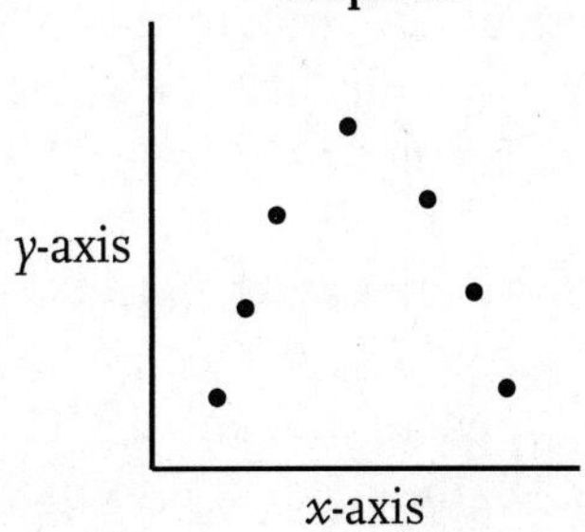

Measures of Average and Dispersion

☞ **Tip #29:** **The mean, median, and mode for a given set of data may all be different values. Data with a high standard deviation is more spread out, and the graph of such data is typically "low" and "wide." Data with a low standard deviation is bunched, and its graph is typically "narrow" and "tall."**

Snapshot

Mean, median, and mode are referred to in statistics as measures of central tendency. They are ways of calculating averages. Mean (also called arithmetic mean or arithmetic average) is calculated as: average = sum ÷ number. We normally calculate average in this way for everyday purposes. Median is the middle-most number, found by taking half the numbers below and half the numbers above, after lining the numbers up in numerical sequence, from lowest to highest (or highest to lowest). Mode is the most frequently recurring number in a set of data.

Let's test the three concepts of central tendency—mean, median, and mode—using a simple example. A science teacher gives his or her class five pop quizzes during the semester, and a given student receives scores (out of a possible 10 points) of: 4.0, 4.0, 5.0, 7.0, and 10.0. What is this student's average score? Well, it depends on which average you are referring to. In this case, the student would have an arithmetic average (mean) of 6.0, a median score of 5.0, and a mode of 4.0. On the other hand, if a student had quiz results of 5.0, 6.0, 9.0, 10.0, and 10.0, then in this case, the student would have an arithmetic average (mean) of 8.0, a median score of 9.0, and a mode of 10.0.

If a student had six pop quizzes during the semester and received scores of 4.0, 4.0, 6.0, 8.0, 10.0, and 10.0, then what would be the median and mode? The arithmetic average (mean) would be 7.0, the median also be 7.0 (average of the middle terms 6.0

and 8.0) and we would have two modes, 4.0 and 10.0. Note that the median is always an average of the two middle terms when we have an even set of data and we never average two or more different modes.

Note that any graph that is symmetrical will have an identical mean, median, and mode. A symmetrical graph is one in which we can draw a line down the middle of that graph, through the graph's high point, and both sides will be mirror images of each other.

Range and standard deviation are measures of dispersion. Range is the difference between the smallest and largest values in a set of data. Standard deviation is a measure of how much variance there is among values in a set of data. In thinking about standard deviation in a broad and conceptual way, we only need remember that the greater the standard deviation, the more the values are dispersed from the mean. A high standard deviation tells us that all or most of the values are spread out; a low standard deviation tells us that all or most of the values are bunched together.

It is true to say that standard deviation is greatest for a heterogeneous group of items and lowest for a homogeneous group of items. A homogeneous group of data is similar in the characteristic we are observing. A heterogeneous group of data is dissimilar in the characteristic we are observing. Say the characteristic we are observing is height. We would expect the standard deviation for the height of a team of professional basketball players to be lower than it would be for a group of people randomly selected at the local mall. Among a team of basketball players, tallness is nothing unusual. In visualizing the standard deviation for a team of such players, think first in terms of the estimated mean (arithmetic) for the height of the players. Once we visualize the mean height, we can see that most players approximate this height, although there certainly are a few shorter players and a few really tall players.

Now suppose we randomly survey the heights of people we find at the local mall. Our sample consists of children, teenagers, and adults across both genders. Obviously, we can expect a far greater standard deviation among individuals selected at random at the mall than the players playing on a professional basketball team. It is equally true to say that the heights of the team players are more homogeneous relative to the heights of people surveyed at the mall, and that the heights of people surveyed at the mall are more heterogeneous relative to the heights of a team of basketball players.

EXHIBIT 4.4 CONSUMER AGE AND PURCHASING PATTERNS

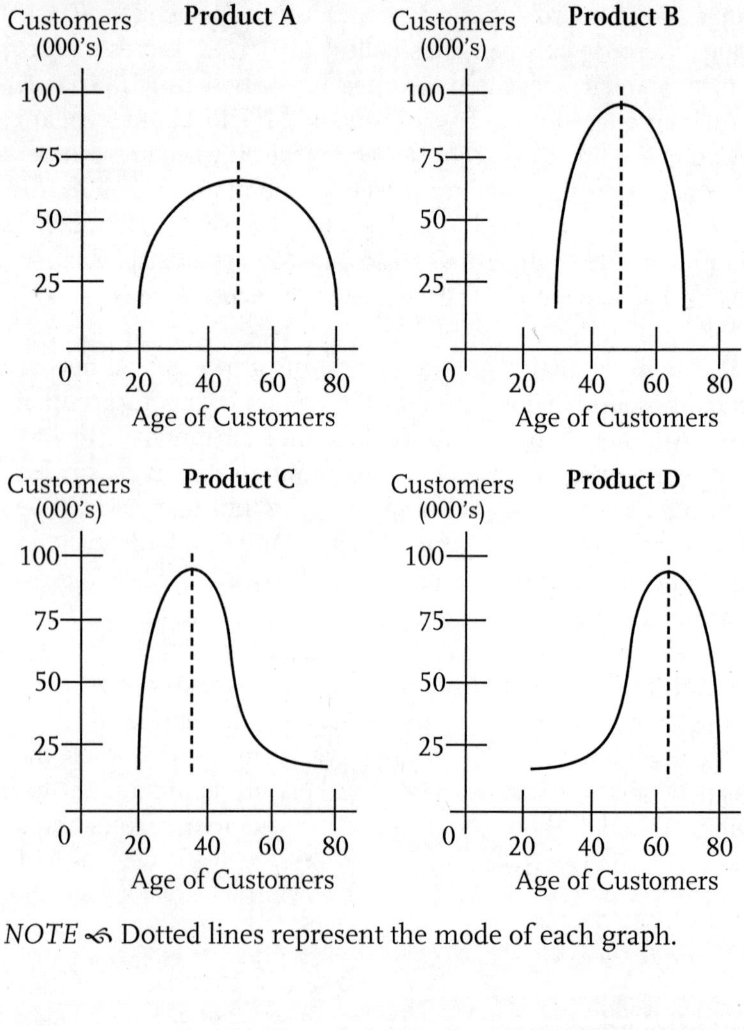

NOTE ⁂ Dotted lines represent the mode of each graph.

The purchase/sales patterns of four different products in a department store were tabulated over a fixed period of time, according to customers' ages, ranging between 20 and 80 years, and the results are displayed in the graphs in *Exhibit 4.4*. Answer the following six questions with respect to the customers' ages.

98. Trends Q1 ( )

The graphs of which individual products show an identical mean, median, and mode?

99. Trends Q2 ( )

The graph of which of the four products shows the smallest mode?

100. Trends Q3 ( )

The graph of which product shows a greater mean compared with its mode?

101. Trends Q4 ( )

The graph of which product shows a greater mode compared with its mean?

102. Trends Q5 ( )

The graph of which product shows the smallest range?

103. Trends Q6 ( )

The graph of which product shows the lowest standard deviation?

Statistical Significance

☞ Tip #30: **Just because two values are different does not mean that they are statistically significantly different. And just because two values are statistically significantly different does not mean that their difference is substantively meaningful.**

Snapshot

A very interesting concept in statistics is the topic of statistical significance: this is the point at which a difference really matters.

Just because two values are different does not mean that they are statistically significantly different.

You are surprised to read that a new study claims that vegetarians live on average up to 18 months longer than do non-vegetarians. A series of questions flood to mind. "Is this difference a big difference? Would I live longer if I became a vegetarian? Isn't living longer also due to chance regardless of one's eating habits?

Statistics—the science of analyzing data and drawing conclusions, often for the purpose of making predictions—comes to the rescue as if to say, "Give me your sample and let's do some calculations, and then we'll be able to tell what this 18-month difference really means." At what point, measured in months or years, does a difference in longevity really matter?

If the difference of 18 months is statistically significant, it would mean that this difference is not due merely to chance but that being a vegetarian really does increase one's longevity. If an 18-month difference is not statistically significant, it would mean that this difference is not enough to link vegetarianism to increased longevity. It could be that chance itself is all that is needed to explain why vegetarians in the study do or do not live a year and a half longer than does the average non-vegetarian.

The phrase "statistical significance" has a special meaning in statistics. It is linked to the figure of 95%. If a difference is statistically significant, it is 95% probable that this difference is genuine and not simply a matter of chance. The 95% level is somewhat mysterious but likely comes from academia, where a theory usually has to have at least a 95% chance of being true to be considered worth telling people about. According to our vegetarian study, if the difference is statistically significant, we could believe with 95% certainty that a switch to vegetarianism would help us increase our longevity on average by one and a half years.

A few additional points. First, in normal English, "significant" means important. In statistics, "significant" means probably true (that is, not due to chance). Statistically significant, therefore, does not necessarily mean important. Again, when a statistic is significant, it doesn't necessarily mean that the finding is important or interesting, or that it has any decision-making utility.

Second, although we have cited the 95% figure, most often, a 5% figure is used in discussing statistical significance. The good news is that these numbers are just mirror opposites of each other. A finding is statistically significant if there is less than a 5% chance that it's not true (or less than a 5% chance that the difference is due to chance). In short, it doesn't matter whether we say that there is a 95% or better chance that something is true or that there is less than a 5% chance that something is not true.

Third, we are proceeding under the assumption that sample sizes derived from studies or surveys are sufficiently large. If a sample size is too small, any claim of a real difference between two values will be inconclusive.

Just because two values are statistically significantly different does not mean that their difference is substantively meaningful.

Even if an 18-month increase in longevity is a statistically significant difference, a person might not conclude that this difference in lifespan is substantively meaningful. That is, a person might not be moved to become a vegetarian because he or she finds the practice too inconvenient and restrictive. For example, it might be difficult finding vegetarian food and/or annoying when ordering in the company of non-vegetarians.

Sometimes, the prospect of adverse consequences must be factored in as a qualitative component of our decision-making process. Say a pharmaceutical company develops a drug to help with cancer, and an outside agency surveys the lifespan of people who do and do not take the drug and concludes that statistics show that the drug enhances lifespan by four weeks. Statistically speaking, that augmented lifespan may be significantly longer for patients who took the drug, but is the difference substantively meaningful? Does that extra four weeks merit the drug's cost and potential side effects? Although statistics may tell us that a given difference matters, it still might not be a meaningful difference, particularly in light of other considerations.

104. Statistician ( )

If the difference between two numbers or values is deemed statistically significant, then which of the following must be true?

I. The difference between these two numbers or values is, by implication, considered important.

II. The difference between these two numbers or values cannot, at the same time, also be considered not statistically significant.

III. The odds that the difference between these two numbers or values cannot be relied upon are less than a 1 in 20 chance.

IV. There is at least a 95% chance that the difference between these two numbers or values can be relied upon, and less than a 5% chance that this difference is due to chance.

A) I & II
B) III & IV
C) I, II & III
D) II, III & IV
E) All of the above

Appendixes

Do not worry about your difficulties in mathematics. I assure you that mine are greater.

—Albert Einstein

Appendix I – Numeracy Principles 1 to 30

Chapter 1 – Basic Numeracy Ingredients

Percentages

☞ Tip #1: A percentage increase is not the same as a percentage of an original number.

☞ Tip #2: You can't add (or subtract) the percentages of different wholes.

☞ Tip #3: Percentages cannot be compared directly to numbers because percentages are relative measures while numbers are actual values.

☞ Tip #4: Percentages (or ratios) are necessary tools for drawing conclusions about qualitative measures, such as likability, preferability, danger, or safety.

☞ Tip #5: Growing pies present a classic case where the actual dollar value of a given item increases over time, yet this same item decreases in percentage terms relative to a larger pie.

☞ Tip #6: Consider the following statement: "Person A spends a greater portion of his or her disposable income on beer than on coffee as compared to person B." This does not necessarily mean that person A spends more money in total on beer and coffee than does person B. It also does not mean that Person A, individually, spends more money on beer than he or she does on coffee.

Ratios & Proportions

☞ Tip #7: Ratios tell us nothing about actual size or value.

☞ Tip #8: You cannot add or subtract a number from a ratio unless you know the exact numbers that make up the ratio. However, you can always multiply or divide both sides of a ratio by a given number.

☞ Tip #9: A part-to-part ratio is not the same thing as a part-to-whole ratio. If the ratio of female to male attendees at a conference is 1:2, this does not mean that 50 percent of the conference attendees are female.

☞ Tip #10: Proportions, in the form of $\frac{A}{B} = \frac{C}{D}$, are among the simplest but most versatile math problem-solving tools.

☞ Tip #11: If product A is selling for 20% more than product B, then the ratio of the selling price of product A to the selling price of product B is 120% to 100% (not 100% to 80%). If product B is selling for 20% less than product A, then the ratio of the selling price of product A to the selling price of B is 100% to 80% (not 120% to 100%).

Reciprocals

☞ Tip #12: Dividing by any number produces the same result as multiplying by that number's reciprocal (and vice-versa).

Order of Operations

☞ Tip #13: The order in which two or more numbers are added or multiplied does not matter. However, the order in which two or more numbers are subtracted or divided does matter.

Chapter 2 – Wonderful Math Recipes

Overlap Scenarios

☞ Tip #14: Situations involving either A or B do not necessarily preclude the possibility of both A and B. When two events overlap, the number of items in both groups will exceed 100% due to mutual inclusivity.

Matrix Scenarios

☞ Tip #15: A matrix is a marvelous tool for summarizing data that is being contrasted across two variables and that can be sorted into four distinct outcomes.

Mixture Scenarios

☞ Tip #16: Mixture problems are best solved using the barrel method, which enables information to be summarized in a 3-row by 3-column "table."

Weighted Average Scenarios

☞ Tip #17: To find a weighted average, multiply each event by its associated weight and total the results. In the case of probabilities, multiply each event by its respective probability and total the results.

Chapter 3 – Favorite Numeracy Dishes

Markup vs. Margin

☞ Tip #18: Markup is always larger than margin.

Price, Cost, Volume, and Profit

☞ Tip #19: An increase (decrease) in sales volume does not necessarily translate to an increase (decrease) in profit.

☞ Tip #20: When price-cost-volume-profit problems are presented in the form of algebraic expressions, it is best to first find a dollar per unit, dollar per person, usage per unit, or usage per person figure.

Break-Even Point

☞ Tip #21: In break-even point problems, break-even occurs exactly where variable revenue (sales revenue less variable costs) equals total fixed costs.

Aggregate Costs vs. Per-Unit Costs

☞ Tip #22: Total costs must be distinguished from per-unit costs, as is the case whenever the number of units varies between two scenarios under comparison.

Efficiency

☞ Tip #23: Efficiency is output divided by input. Cost efficiency is cost divided by efficiency.

Distribution and Allocation

☞ Tip #24: Elements within distributions may be uniformly spaced or concentrated in "bunches." Allocations may be equal or unequal, proportionate or disproportionate.

☞ Tip #25: Front-loading occurs when a greater proportion of inputs (for example, time, materials, costs, efforts, benefits) are entered early in a process. Back-loading occurs when a greater proportion of inputs are entered late in a process.

Chapter 4 – Special Math Garnishments

Translating Simple Graphs

☞ Tip #26: Forward-slashing graphs indicate that two variables are directly related. Back-slashing graphs indicate that two variables are indirectly (inversely) related. Graphs formed with straight lines have variables which are linearly related. Graphs formed with curved lines have variables that are non-linearly related.

Line Graphs, Pie Charts, and Bar Charts

☞ Tip #27: Line graphs are best used to show how data changes over time. Pie charts are great for breaking data down by category. Bar charts are ideal for side-by-side comparisons of up to six pieces of data. Knowing the strength of each type of graph or chart and how they are best used to present information also helps us in interpreting the underlying data.

Correlation

☞ Tip #28: Positive correlations have graphs sloping upward; negative correlations have graphs sloping downward. The stronger the correlation between two variables, the nearer to a straight line will be their graph; the graph of a perfect correlation is a straight line.

Measures of Average and Dispersion

☞ Tip #29: The mean, median, and mode for a given set of data may be all different values. Data with a high standard deviation is more spread out, and the graph of such data is typically "low" and "wide." Data with a low standard deviation is bunched, and its graph is typically "narrow" and "tall."

Statistical Significance

☞ Tip #30: Just because two values are different does not mean that they are statistically significantly different. And just because two values are statistically significantly different does not mean that their difference is substantively meaningful.

Appendix II – The World of Numbers

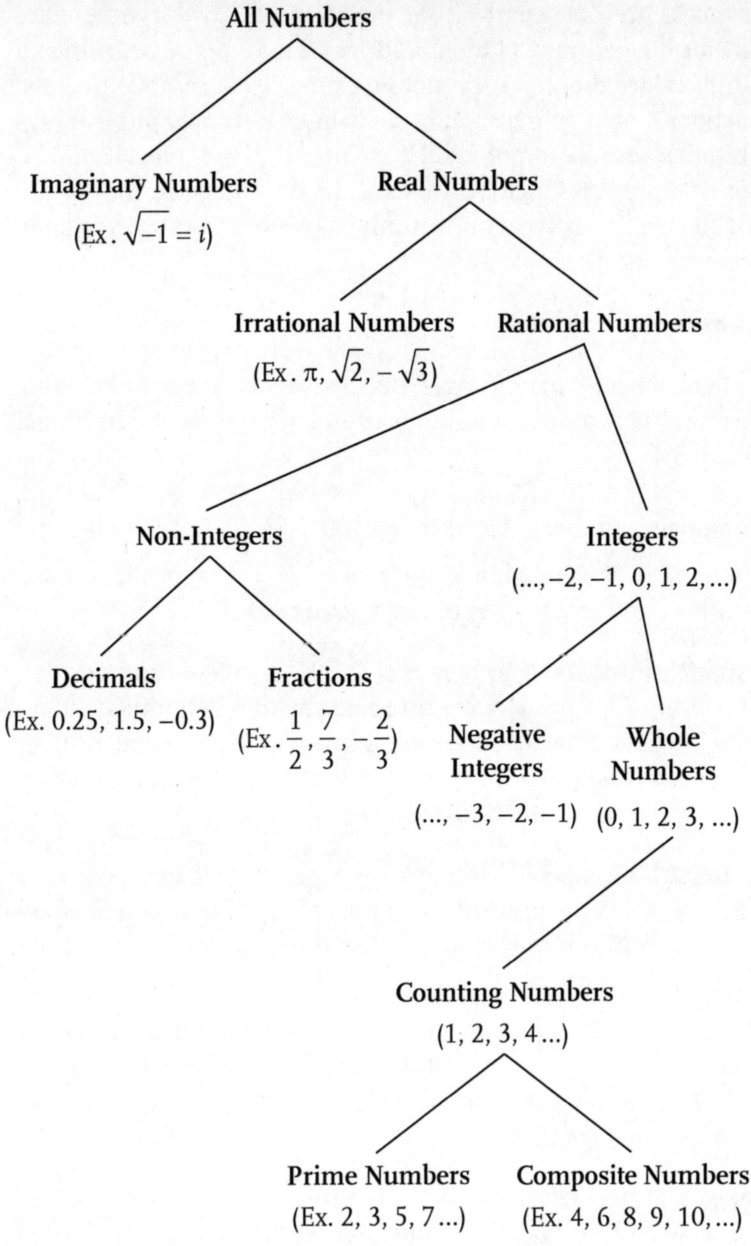

With reference to the flowchart on the previous page, numbers are first divided into real and imaginary numbers. *Imaginary* numbers are not a part of our everyday life. *Real* numbers are further divided into rational and irrational numbers. *Irrational* numbers are those that cannot be expressed as simple integers, fractions, or decimals; non-repeating decimals are always irrational numbers, of which π may be the most famous. *Rational* numbers include integers and non-integers. *Integers* are positive and negative whole numbers, while *non-integers* include decimals and fractions.

Number Definitions

Real numbers: Any number that exists on the number line. Real numbers are the combined group of rational and irrational numbers.

Imaginary numbers: Any number multiplied by *i*, the imaginary unit: $i = \sqrt{-1}$. Imaginary numbers are the opposite of real numbers and are not part of our everyday life.

Rational numbers: Numbers that can be expressed as a fraction whose top (the numerator) and bottom (the denominator) are both integers. Note that every whole number is a rational number because any whole number can be written as a fraction (that is, any number can itself be placed over the number 1).

Irrational numbers: Numbers that can't be expressed as a fraction or, more precisely, cannot be expressed as a fraction whose top (the numerator) and bottom (the denominator) are integers. Square roots of non-perfect squares (such as $\sqrt{2}$) are irrational, and π is irrational. An irrational number may be described as a non-repeating decimal or, stated another way, an irrational number has endless non-repeating digits to the right of its decimal point.

Integers: Integers consist of those numbers that are multiples of 1: $\{\ldots, -2, -1, 0, 1, 2, 3, \ldots\}$. Integers are "integral": they contain no fractional or decimal parts.

Non-integers: Non-integers are those numbers that contain fractional or decimal parts. For example, $\frac{1}{2}$ and 0.125 are non-integers.

Whole numbers: Non-negative integers: $\{0, 1, 2, 3, \ldots\}$. Note that 0 is a whole number but it is technically a non-negative whole number, given that it is neither positive nor negative).

Counting numbers: The subset of whole numbers which excludes 0: $\{1, 2, 3, \ldots\}$.

Prime numbers: Prime numbers are a subset of the counting numbers. They include those non-negative integers that have two and only two factors; that is, the factors 1 and themselves. The first 10 primes are 2, 3, 5, 7, 11, 13, 17, 19, 23, and 29. Note that 1 is not a prime number as it only has one factor, that is, 1. Also, the number 2 is not only the smallest prime but also the only even prime number.

Composite numbers: A positive number that has more than two factors other than 1 and itself. Also, any non-prime number greater than 1. Examples include: 4, 6, 8, 9, and 10. Note that 1 is not a composite number, and the number 4 is the smallest composite number.

Fraction: A fraction is a mathematical term that is usually expressed in the form $\frac{a}{b}$, where a is the numerator and b is the denominator. Although most everyday fractions are between 0 and positive 1, fractions can be greater than 1 and they can also be negative.

Decimal: Decimal means "based on ten." Each place or position left of the decimal point is 10 times bigger than the previous place or position. Each place or position right of the decimal point is 10 times smaller than the previous place or position. The exact opposite is true in the case of negative numbers. Whole numbers are typically written without showing a decimal place.

The Four Basic Operations

The four basic arithmetic operations are addition, subtraction, multiplication, and division. The results of these operations are called sum, difference, product, and quotient, respectively. Two additional operations involve exponents and radicals.

Addition

2	← augend
+3	← addend
5	← sum

Subtraction

25	← minuend
−16	← subtrahend
9	← difference

Multiplication

3	← multiplicand
×4	← multiplier
12	← product

Division

10	← dividend
÷5	← divisor
2	← quotient

Exponents

2^3	← power (3), ← base (2)
$= 8$	← result

Radicals

$\sqrt{9}$	← radical ($\sqrt{\ }$), ← radicand (9)
$= 3$	← root

Common Fractions and their Percentage Equivalents

Exercise – Fill in the missing percentages to change common fractions into their percentage equivalents.

$\frac{1}{1} = 100\%$

$\frac{1}{2} = 50\%$ $\frac{2}{2} = 100\%$

$\frac{1}{3} = ?\%$ $\frac{2}{3} = ?\%$ $\frac{3}{3} = 100\%$

$\frac{1}{4} = ?\%$ $\frac{2}{4} = 50\%$ $\frac{3}{4} = ?\%$ $\frac{4}{4} = 100\%$

$\frac{1}{5} = ?\%$ $\frac{2}{5} = ?\%$ $\frac{3}{5} = ?\%$ $\frac{4}{5} = ?\%$ $\frac{5}{5} = 100\%$

$\frac{1}{6} = ?\%$ $\frac{2}{6} = ?\%$ $\frac{3}{6} = 50\%$ $\frac{4}{6} = ?\%$ $\frac{5}{6} = ?\%$ $\frac{6}{6} = 100\%$

$\frac{1}{7} = ?\%$ $\frac{2}{7} = ?\%$ $\frac{3}{7} = ?\%$ $\frac{4}{7} = ?\%$ $\frac{5}{7} = ?\%$ $\frac{6}{7} = ?\%$ $\frac{7}{7} = 100\%$

$\frac{1}{8} = ?\%$ $\frac{2}{8} = ?\%$ $\frac{3}{8} = ?\%$ $\frac{4}{8} = 50\%$ $\frac{5}{8} = ?\%$ $\frac{6}{8} = ?\%$ $\frac{7}{8} = ?\%$ $\frac{8}{8} = 100\%$

$\frac{1}{9} = ?\%$ $\frac{2}{9} = ?\%$ $\frac{3}{9} = ?\%$ $\frac{4}{9} = ?\%$ $\frac{5}{9} = ?\%$ $\frac{6}{9} = ?\%$ $\frac{7}{9} = ?\%$ $\frac{8}{9} = ?\%$ $\frac{9}{9} = 100\%$

$\frac{1}{10} = ?\%$ $\frac{2}{10} = ?$ $\frac{3}{10} = ?\%$ $\frac{4}{10} = ?$ $\frac{5}{10} = 50\%$ $\frac{6}{10} = ?$ $\frac{7}{10} = ?$ $\frac{8}{10} = ?$ $\frac{9}{10} = ?$ $\frac{10}{10} = 100\%$

Solutions – Common fractions and their percentage equivalents.

$\frac{1}{1} = 100\%$

$\frac{1}{2} = 50\%$ $\frac{2}{2} = 100\%$

$\frac{1}{3} = 33\frac{1}{3}\%$ $\frac{2}{3} = 66\frac{2}{3}\%$ $\frac{3}{3} = 100\%$

$\frac{1}{4} = 25\%$ $\frac{2}{4} = 50\%$ $\frac{3}{4} = 75\%$ $\frac{4}{4} = 100\%$

$\frac{1}{5} = 20\%$ $\frac{2}{5} = 40\%$ $\frac{3}{5} = 60\%$ $\frac{4}{5} = 80\%$ $\frac{5}{5} = 100\%$

$\frac{1}{6} = 16\frac{2}{3}\%$ $\frac{2}{6} = 33\frac{1}{3}\%$ $\frac{3}{6} = 50\%$ $\frac{4}{6} = 66\frac{2}{3}\%$ $\frac{5}{6} = 83\frac{1}{3}\%$ $\frac{6}{6} = 100\%$

$\frac{1}{7} = 14.3\%$ $\frac{2}{7} = 28.6\%$ $\frac{3}{7} = 42.9\%$ $\frac{4}{7} = 57.1\%$ $\frac{5}{7} = 71.4\%$ $\frac{6}{7} = 85.7\%$ $\frac{7}{7} = 100\%$

$\frac{1}{8} = 12.5\%$ $\frac{2}{8} = 25\%$ $\frac{3}{8} = 37.5\%$ $\frac{4}{8} = 50\%$ $\frac{5}{8} = 62.5\%$ $\frac{6}{8} = 75\%$ $\frac{7}{8} = 87.5\%$ $\frac{8}{8} = 100\%$

$\frac{1}{9} = 11.11\%$ $\frac{2}{9} = 22.22\%$ $\frac{3}{9} = 33.33\%$ $\frac{4}{9} = 44.44\%$ $\frac{5}{9} = 55.55\%$ $\frac{6}{9} = 66.66\%$ $\frac{7}{9} = 77.77\%$ $\frac{8}{9} = 88.88\%$ $\frac{9}{9} = 100\%$

$\frac{1}{10} = 10\%$ $\frac{2}{10} = 20\%$ $\frac{3}{10} = 30\%$ $\frac{4}{10} = 40\%$ $\frac{5}{10} = 50\%$ $\frac{6}{10} = 60\%$ $\frac{7}{10} = 70\%$ $\frac{8}{10} = 80\%$ $\frac{9}{10} = 90\%$ $\frac{10}{10} = 100\%$

Rules for Odd and Even Numbers

Even + Even = Even

Example $2 + 2 = 4$
Example $-2 + -2 = -4$

Odd + Odd = Even

Example $3 + 3 = 6$
Example $-3 + -3 = -6$

Even + Odd = Odd

Example $2 + 3 = 5$
Example $-2 + -3 = -5$

Odd + Even = Odd

Example $3 + 2 = 5$
Example $-3 + -2 = -5$

Even – Even = Even

Example $4 - 2 = 2$
Example $-4 - (-2) = -2$

Odd – Odd = Even

Example $5 - 3 = 2$
Example $-5 - (-3) = -2$

Even – Odd = Odd

Example $6 - 3 = 3$
Example $-6 - (-3) = -3$

Odd – Even = Odd

Example $5 - 2 = 3$
Example $-5 - (-2) = -3$

Even × Even = Even

Example $2 \times 2 = 4$
Example $-2 \times -2 = 4$

Odd × Odd = Odd

Example $3 \times 3 = 9$
Example $-3 \times -3 = 9$

Even × Odd = Even

Example $2 \times 3 = 6$
Example $-2 \times -3 = 6$

Odd × Even = Even

Example $3 \times 2 = 6$
Example $-3 \times -2 = 6$

Even ÷ Even = Even

Example $4 \div 2 = 2$
Example $-4 \div -2 = 2$

Odd ÷ Odd = Odd

Example $9 \div 3 = 3$
Example $-9 \div -3 = 3$

Even ÷ Odd = Even

Example $6 \div 3 = 2$
Example $-6 \div -3 = 2$

Odd ÷ Even = Not an Integer

Example $5 \div 2 = 2\frac{1}{2}$
Example $-5 \div -2 = 2\frac{1}{2}$

With reference to the previous two examples, an odd number divided by an even number does not result in either an even or odd integer; it results in a non-integer.

Rules for Positive and Negative Numbers

Positive + Positive = Positive

Example $2 + 2 = 4$

Negative + Negative = Negative

Example $-2 + (-2) = -4$

Positive + Negative = Depends

Example $4 + (-2) = 2$
Example $2 + (-4) = -2$

Negative + Positive = Depends

Example $-2 + 4 = 2$
Example $-4 + 2 = -2$

Positive – Positive = Depends

Example $4 - 2 = 2$
Example $2 - 4 = -2$

Negative – Negative = Depends

Example $-2 - (-4) = 2$
Example $-4 - (-2) = -2$

Positive – Negative = Positive

Example $2 - (-2) = 4$

Negative – Positive = Negative

Example $-4 - 2 = -6$

Positive × Positive = Positive

Example $2 \times 2 = 4$

Negative × Negative = Positive

Example $-2 \times -2 = 4$

Positive × Negative = Negative

Example $2 \times -2 = -4$

Negative × Positive = Negative

Example $-2 \times 2 = -4$

Positive ÷ Positive = Positive

Example $4 \div 2 = 2$

Negative ÷ Negative = Positive

Example $-4 \div -2 = 2$

Positive ÷ Negative = Negative

Example $4 \div -2 = -2$

Negative ÷ Positive = Negative

Example $-4 \div 2 = -2$

Common Squares, Cubes, and Square Roots

Basic Squares from 1 to 30			Basic Cubes from 1 to 10
$1^2 = 1$	$11^2 = 121$	$21^2 = 441$	$1^3 = 1$
$2^2 = 4$	$12^2 = 144$	$22^2 = 484$	$2^3 = 8$
$3^2 = 9$	$13^2 = 169$	$23^2 = 529$	$3^3 = 27$
$4^2 = 16$	$14^2 = 196$	$24^2 = 576$	$4^3 = 64$
$5^2 = 25$	$15^2 = 225$	$25^2 = 625$	$5^3 = 125$
$6^2 = 36$	$16^2 = 256$	$26^2 = 676$	$6^3 = 216$
$7^2 = 49$	$17^2 = 289$	$27^2 = 729$	$7^3 = 343$
$8^2 = 64$	$18^2 = 324$	$28^2 = 784$	$8^3 = 512$
$9^2 = 81$	$19^2 = 361$	$29^2 = 841$	$9^3 = 729$
$10^2 = 100$	$20^2 = 400$	$30^2 = 900$	$10^3 = 1000$

Also:

$10^1 = 10$
$10^2 = 100$
$10^3 = 1,000$
$10^4 = 10,000$
$10^5 = 100,000$
$10^6 = 1,000,000$
$10^9 = 1$ billion
$10^{12} = 1$ trillion

NOTE ↜ In most English-speaking countries today (particularly the U.S., Great Britain, Canada, and Australia), one billion equals 1,000,000,000 or 10^9, or one thousand millions. In many other countries, including France, Germany, Spain, Norway, and Sweden, the word "billion" denotes 10^{12}, or one million millions. Although Britain and Australia have traditionally employed the international usage of 10^{12}, they have now largely switched to the U.S. usage of 10^9.

Common Square Roots

$\sqrt{1} = 1$

$\sqrt{2} = 1.4$

$\sqrt{3} = 1.7$

$\sqrt{4} = 2$

$\sqrt{5} = 2.2$

Pop Quiz

See pages 258–259 for answers.

105. Fractions to Percents ( )

Convert the following fractions to their percentage equivalents:

$\frac{1}{3} =$ $\frac{2}{3} =$ $\frac{1}{6} =$ $\frac{5}{6} =$ $\frac{1}{8} =$

$\frac{3}{8} =$ $\frac{5}{8} =$ $\frac{7}{8} =$ $\frac{1}{9} =$ $\frac{5}{9} =$

106. Decimals to Fractions ( )

Convert the following decimals, which are greater than 1, into fractional equivalents.

1.2 = 1.25 = 1.33 =

Simplify each expression below without multiplying decimals or using a calculator. (Hint: Translate each decimal into a common fraction and then multiply and/or divide these fractions.)

$$(1.25)(0.50)(0.80)(2.00) =$$

$$\frac{(0.7500)(0.8333)}{(0.6250)} =$$

$$\frac{(0.2222)}{(0.3333)(0.6666)} =$$

107. Common Square Roots ( )

Put the following statements in order, from largest to smallest value. (Hint: Approximate each square root to one decimal point.)

I. $1 + \sqrt{5}$

II. $2 + \sqrt{3}$

III. $3 + \sqrt{2}$

Divisibility Rules

No.	Divisibility Rule	Examples
1	Every number is divisible by 1.	15 divided by 1 equals 15.
2	A number is divisible by 2 if it is even.	24 divided by 2 equals 12.
3	A number is divisible by 3 if the sum of its digits is divisible by 3.	651 is divisible by 3 since 6 + 5 + 1 = 12 and "12" is divisible by 3.
4	A number is divisible by 4 if its last two digits form a number that is divisible by 4.	1,112 is divisible by 4 since the number "12" is divisible by 4.
5	A number is divisible by 5 if the number ends in 5 or 0.	245 is divisible by 5 since this number ends in 5.
6	A number is divisible by 6 if it is divisible by both 2 and 3.	738 is divisible by 6 since this number is divisible by both 2 and 3, and the rules that govern the divisibility of 2 and 3 apply.
7	No clear rule.	Not applicable.
8	A number is divisible by 8 if its last three digits form a number that is divisible by 8.	2,104 is divisible by 8 since the number "104" is divisible by 8.
9	A number is divisible by 9 if the sum of its digits is divisible by 9.	4,887 is divisible by 9 since 4 + 8 + 8 + 7 = 27 and 27 is divisible by 9.
10	A number is divisible by 10 if it ends in 0.	990 is divisible by 10 because 990 ends in 0.

Appendix III – Translating Words into Numbers

Classic Translation Scenarios

In order to solve basic math problems, we need to be able to do two things. First, we need to *translate* "words" into proper math. Second, we need to *solve* problems accurately.

"Age" problems provide a truly classic example of the process by which we translate words into math and, therefore, are a great starting point for reviewing translation skills.

Here's an overview of four tricky areas of age translations:

I. i) Bill is 9 years older than Alice.

$B = A + 9$

ii) Alice is 9 years younger than Bill.

$A = B - 9$

II. i) Dorothy is twice as old as Caren.

$D = 2C$

ii) Caren is half as old as Dorothy.

$C = \frac{1}{2}D$

III. i) Five years from now, Joe will be twice as old as Hugh.

$J + 5 = 2(H + 5)$

ii) Five years from now, Hugh will be half as old as Joe.

$\frac{1}{2}(J + 5) = H + 5$

Chart of Basic Math Translations

English		Math	
Term	**Example**	**Symbol**	**Example**
add	2 added to 3	+	2 + 3
sum	sum of 2 and 3	+	2 + 3
plus	2 plus 3	+	2 + 3
more	3 more than 2	+	2 + 3
subtract	5 subtract 2	–	5 – 2
minus	5 minus 2	–	5 – 2
less	5 less 2	–	5 – 2
difference	difference between 5 and 2	–	5 – 2
multiply	2 multiplied by 3	×	2 × 3
product	product of 2 and 3	×	2 × 3
divide	6 divided by 3	÷	6 ÷ 3
per	8 miles per hour	/	8/hour
equal	3 is equal to 3	=	3 = 3
result	... results in 3	=	= 3
greater than	3 is greater than 2	>	3 > 2
less than	2 less than 3	<	2 < 3

IV. i) Ten years from now, Sam will be twice Tina's current age.

$S + 10 = 2T$

ii) Ten years from now, Tina will be half Sam's current age.

$\frac{1}{2}T = S + 10$

Here are some follow up notes to clarify the above translations.

I. In cases where an age problem states, "A is 10 years older than B," the math translation is $A = B + 10$. Remember that the 10 years is added to B's side of the equation in order to balance the equation. This may seem counterintuitive, but since we are dealing with an equation (note the "equals" sign), we must add 10 to B's side of the equation, and not add 10 to A's side of the equation.

II. In cases where an age problem states, "A is twice as old as B," the math translation is $A = 2B$, not $2A = B$. Again, this may seem counterintuitive, but since we are dealing with an equation, we must multiply B's side of the equation by 2, not the other way around.

III. In cases where an age problem states, "10 years from now A will be twice as old as B," the math translation is $A + 10 = 2(B + 10)$. Do not write $A + 10 = 2B$. In this situation, remember to add 10 years to both sides of the equation because both individuals will have aged 10 years.

IV. Contrast the statements: "10 years from now, A will be twice as old as B," and "10 years from now, A will be twice as old as B is now." The first statement (per scenario III above) requires that 10 years be added to both sides. The second statement (per this scenario IV) requires that 10 years be added only to one side of the equation.

108. Leo (🌶)

Elmer, the circus Elephant, is currently ten years older than Leo, the circus Lion. In five years from now, Elmer, the circus Elephant, will be exactly twice as old as Leo, the circus Lion. How old is Leo today?

A) 3
B) 5
C) 15
D) 22
E) 25

109. Brad and Janet (🌶🌶)

If Janet were eight years younger, she would be half as old as Brad is now. If Janet were two years older, she would be that same age as Brad is now. How old is Janet?

A) 8
B) 16
C) 18
D) 24
E) 28

How to Solve Equations

This book deals with arithmetic and basic algebra. What is the difference between arithmetic and algebra? Arithmetic concerns itself with numbers, factions, decimals and percents, and the basic operations of addition, subtraction, multiplication, and division. Algebra deals with symbols, usually in the form of variables, to express relationships among these variables. As a practical matter, whenever we are dealing with equations, we are typically dealing with algebra.

Let's get some basic definitions out of the way.

Equation: An equation consists of one or more mathematical terms that are joined by an equal sign (=).

Variable: A variable is some number that stands for another number (for example, x, y, q, E). It is common practice to italicize variables when they appear in print.

Formula: A formula is a special kind of equation that shows a relationship among different variables. Ex. $A = l \times w$. This may also be written out as *Area* = *length* × *width*.

How do we name variables? With respect to "naming" variables, sometimes we choose a letter, such as the first letter of the word that the variable stands for, whether capitalized or not. Other times, we choose to use traditional letters such as x and y. There are no hard-and-fast rules. It is important to remain flexible and be able to work with variables however they are presented.

How do we solve equations? We will likely encounter one of two situations. If we need to solve for one variable within one equation, it is simply a matter isolating the single variable on one side of the equation (usually the left-hand side) while placing all other terms on the right-hand side. If we need to solve using two variables (and we also have two distinct equations), we will typically substitute for one of the variables. Let's review how this works.

Consider the following problem:

> A student took a true-false math quiz that contained 100 questions and got three questions correct for every one wrong answer. How many questions did the student get correct?

This is an interesting problem that lends itself to being solved with either a single equation (and single variable) or two equations (each containing the same two variables).

Solving this problem using a single equation (and a single variable).

To solve this problem with a single variable, we will let the variable w represent the number of questions that the student got incorrect (wrong answers). This means that $3w$ represents the number of questions the student got correct. Also, we know that there were 100 questions and, logically, that the number of correct answers plus the number of incorrect questions must equal 100 questions.

$$3w + w = 100$$
$$4w = 100$$
$$w = 25$$

Since the number of incorrect questions is 25, the number of correct questions is $3w$ or $3(25)$. The student got 75 questions correct.

Solving this problem using two equations (and two variables).

To solve this problem with two variables, we need two equations. If w represents the number of incorrect questions (wrong answers), then r represents the number of correct questions (right answers). Two equations arise. The first equation below is derived from the fact that the number of correct answers and incorrect answers must total 100. The second equation below is derived from the fact that the number of correct answers (r) must mathematically equal three times the number of incorrect answers ($3w$).

1) $r + w = 100$
2) $r = 3w$

Since the second equation above states that $r = 3w$, we can substitute for the variable r in the first equation as follows:

$$3w + w = 100$$
$$4w = 100$$
$$w = 25$$

Again, since the number of incorrect questions is 25, the number of correct questions is 75, as derived from $3w$ or $3(25)$. Actually, once we have solved for one variable, we can substitute that variable into either of the two equations, and preferably the "easier" of the two.

110. Hardware ( )

Hammers and wrenches are manufactured at a uniform weight per hammer and a uniform weight per wrench. If the total weight of two hammers and three wrenches is one-third that of 8 hammers and 5 wrenches, then the total weight of one wrench is how many times that of one hammer?

A) $\frac{1}{2}$ B) $\frac{2}{3}$ C) 1 D) $\frac{3}{2}$ E) 2

111. Snooker ( )

A snooker tournament charges \$45.00 for VIP seats and \$15.00 for general admission ("regular") seats. On a certain night, a total of 320 tickets were sold, for a total cost of \$7,500. How many general admission seat tickets were sold?

A) 70
B) 90
C) 140
D) 230
E) 250

112. Chili Paste ()

Each week a restaurant serving Mexican food orders the same volume of chili paste, which comes in either 25-ounce cans or 15-ounce cans. If the restaurant must order 40 more of the smaller cans than the larger cans to fulfill its weekly needs, then how many *smaller* cans are required to fulfill its weekly needs?

A) 60
B) 70
C) 80
D) 100
E) 120

Making Tricky Conversions

This segment deals with tricky math conversions and focuses on a very common type of word problem known as "distance-rate-time" problems or D-R-T for short. In particular, we will focus on conversions based on time. Depending on the situation, we will be multiplying by 60 or dividing by 60. It's a no-brainer that there are 60 minutes in an hour, and 60 seconds in a minute. However, sometimes a conversion may seem counterintuitive. The problems presented provide a great mental workout!

Let's start off with a review of the three classic distance-rate-time formulas:

$D = R \times T$ [Distance equals rate times time, where D is distance, R is rate, and T is time.]

$R = \frac{D}{T}$ [Rate equals distance overtime. The terms "rate" and "speed" are interchangeable.]

$T = \frac{D}{R}$ [Time equals distance over rate.]

Let's use very simple examples. We'll use a three-step approach to solve these problems: i) identify the correct formula, in this case $D = R \times T$, ii) fill in the known values, and iii) solve the equation.

1) How far in miles was a trip if the driver drove at 60 miles per hour for 2 hours?

Here we solve for distance.

i) $D = R \times T$

ii) $D = 60 \frac{\text{miles}}{\cancel{\text{hour}}} \times 2 \cancel{\text{hours}}$

iii) $D = 120 \text{ miles}$

2) What was the average speed for a trip of 120 miles that took 2 hours to drive?

Here we solve for rate.

i) $D = R \times T$

ii) $120 \text{ miles} = R \times 2 \text{ hours}$

iii) $R = \frac{120 \text{ miles}}{2 \text{ hours}}$

$R = 60 \text{ miles per hour}$

3) How long would it take to drive 120 miles at an average speed of 60 miles per hour?

Here we solve for time.

i) $D = R \times T$

ii) $120 \text{ miles} = 60 \text{ miles per hour} \times T$

iii) $T = \frac{120 \text{ miles}}{\frac{60 \text{ miles}}{\text{hour}}}$

$$T = 120 \text{ \sout{miles}} \times \frac{1\text{ hour}}{60 \text{ \sout{miles}}}$$

$$T = 2 \text{ hours}$$

In each of the problems that follow in this section, we need to convert units—from minutes to hours or hours to minutes. A key step is to identify what units you need to end up with and which units need to be canceled. The fraction that is multiplied to convert the units will have the number 1 in either the numerator or denominator. The way in which units conveniently cancel when multiplied by this fraction is know in math as "the magic of multiplying by 1."

We will now use a four-step approach to solve these problems: i) identify the correct formula, in this case $D = R \times T$, ii) fill in the known values, iii) convert the units, and iv) solve the equation.

Note: The following three problems need to be converted from minutes to hours.

Distance:

1) A driver drives his car at a rate of 1 mile per minute. How many miles does he drive in x hours?

   i) $D = R \times T$

   ii) $D = \frac{1 \text{ mile}}{\text{minute}} \times x \text{ hours}$

   iii) $D = \frac{1 \text{ mile}}{\text{\sout{minute}}} \times x \text{ \sout{hours}} \times \frac{60 \text{ \sout{minutes}}}{1 \text{ \sout{hour}}}$

   iv) $D = 60x$ miles

Rate:

2) A driver drives 120 miles in 120 minutes. What was his average speed in miles per hour?

i) $D = R \times T$

ii) $120 \text{ miles} = R \times 120 \text{ minutes}$

$$R = \frac{120 \text{ miles}}{120 \text{ minutes}}$$

iii) $R = \dfrac{120 \text{ miles}}{120 \cancel{\text{minutes}}} \times \dfrac{60 \cancel{\text{minutes}}}{1 \text{ hour}}$

iv) $R = 60$ miles per hour

Time:

3) A driver completes a journey of 120 miles. He traveled at a rate of 1 mile per minute. How many <u>hours</u> did the journey take?

i) $D = R \times T$

ii) $120 \text{ miles} = 1 \text{ mile per minute} \times T$

$$T = \frac{120 \text{ miles}}{\dfrac{1 \text{ mile}}{\text{minute}}}$$

iii) $T = 120 \cancel{\text{miles}} \times \dfrac{1 \cancel{\text{minute}}}{1 \cancel{\text{mile}}} \times \dfrac{1 \text{ hour}}{60 \cancel{\text{minutes}}}$

iv) $T = 2$ hours

Note: The following three problems require conversions from <u>hours</u> to <u>minutes</u>.

Distance:

1) A driver drives her car at a rate of 60 miles per hour. How many <u>miles</u> does she drive in x minutes?

i) $D = R \times T$

ii) $D = \dfrac{60 \text{ miles}}{\text{hour}} \times x \text{ minutes}$

iii) $D = \frac{60 \text{ miles}}{\cancel{\text{hour}}} \times x \cancel{\text{ minutes}} \times \frac{1 \cancel{\text{ hour}}}{60 \cancel{\text{ minutes}}}$

iv) $D = 60x$ miles

Rate:

2) A driver drives 120 miles in 2 hours. What was her average speed in miles per minute?

i) $D = R \times T$

ii) 120 miles $= R \times 2$ hours

$$R = \frac{120 \text{ miles}}{2 \text{ hours}}$$

iii) $R = \frac{120 \text{ miles}}{2 \cancel{\text{ hours}}} \times \frac{1 \cancel{\text{ hour}}}{60 \text{ minutes}}$

iv) $R = 1$ mile per minute

Time:

3) A driver completes a journey of 120 miles. She traveled at a rate of 60 miles per hour. How many minutes did the journey take?

i) $D = R \times T$

ii) 120 miles = 60 miles per hour $\times T$

$$T = \frac{120 \text{ miles}}{\frac{60 \text{ miles}}{\text{hour}}}$$

iii) $T = 120 \cancel{\text{ miles}} \times \frac{1 \cancel{\text{ hour}}}{60 \cancel{\text{ miles}}} \times \frac{60 \text{ minutes}}{1 \cancel{\text{ hour}}}$

iv) $T = 120$ minutes

113. Bullet Train #1 ( )

At the rate of d miles per q minutes, how many miles does a bullet train travel in x hours?

A) $\frac{dx}{q}$ B) $\frac{60dx}{q}$ C) $\frac{dx}{60q}$ D) $\frac{60qx}{d}$ E) $\frac{qx}{60d}$

114. Bullet Train #2 ( )

A bullet train completes a journey of d miles. If the journey took q minutes, what was the train's speed in miles per hour?

A) $\frac{d}{q}$ B) $\frac{60q}{d}$ C) $\frac{60d}{q}$ D) $\frac{d}{60q}$ E) $\frac{q}{60d}$

115. Bullet Train #3 ( )

Another bullet train completes a journey of d miles. If this train traveled at a rate of z miles per minute, how many hours did the journey take?

A) $\frac{d}{z}$ B) $\frac{z}{60d}$ C) $\frac{60d}{z}$ D) $\frac{d}{60z}$ E) $\frac{60z}{d}$

116. Run-Run ( )

If Susan takes 9 seconds to run y yards, how many minutes will it take her to run x yards at the same rate?

A) $\frac{xy}{9}$ B) $\frac{9x}{60y}$ C) $\frac{60xy}{9}$ D) $\frac{xy}{540}$ E) $\frac{540x}{y}$

Appendix IV – Tackling Squares and Cubes

A square is a number that results when a given number is multiplied by itself. A cube is a number that results when a given number is multiplied by itself three times. Thus, the square of 5 is 25 because $5 \times 5 = 25$. Likewise, the cube of 5 is 125 because $5 \times 5 \times 5 = 125$. In this segment, we are interested in how numbers behave when multiplied by themselves one or more times. Specifically, we will look at what happens when numbers are squared or cubed. The "behavior" we will observe is whether they become smaller or larger in size.

We are interested here in real numbers, so it is natural to look at the number line. And when we do, something interesting is revealed. Although there are an infinite number of real numbers, there are only seven we need to concern ourselves with in order to draw conclusions about the behavior of all numbers. Again, we are only concerned with what happens when numbers are squared or cubed and whether they become larger or smaller numbers as a result of being squared or cubed.

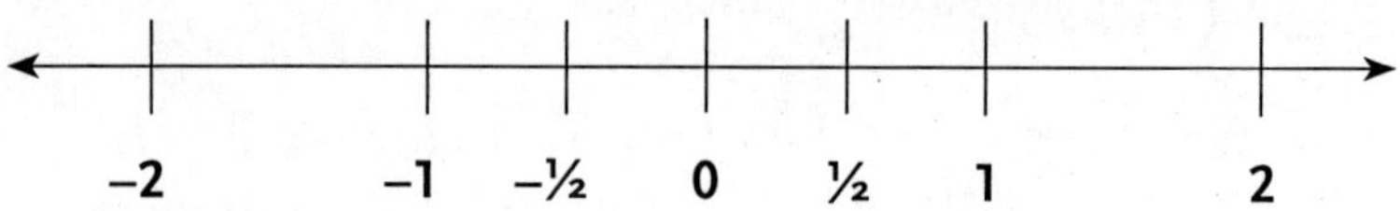

The seven numbers shown above are referred to in this material as the "big seven" numbers (that is, –2, –1, $-\frac{1}{2}$, 0, $\frac{1}{2}$, 1, 2). They are representative of a huge cross section of numbers and actually embody the basic properties of all numbers. Think of each big seven number as a representative of a larger group of numbers. The way the number 2 behaves when squared or cubed is the same way any number greater than 1 behaves when squared or cubed. So the number 2 is our representative for numbers greater than 1 and, for simplicity, we can ignore other numbers such as 3 and 4. The way fractions such as $\frac{1}{2}$ behave when squared or cubed is identical to the way in which fractions between 0 and 1 behave when squared or cubed. In this respect, we can ignore fractions such as $\frac{1}{4}$ or $\frac{3}{4}$, because $\frac{1}{2}$ is our "big seven" representative for fractions between 0 and 1.

Here is a very useful summary of how each of the "big seven" numbers behaves when squared or cubed:

The number 2:

Squaring a number greater than 1 (for example, 2) results in a larger number.

Example: $2^2 = 2 \times 2 = 4$

Result: 4 is bigger than 2

Cubing a number greater than 1 (for example, 2) results in a larger number.

Example: $2^3 = 2 \times 2 \times 2 = 8$

Result: 8 is bigger than 2

The number 1:

Squaring or cubing the number 1 results in the same number.

Examples: $1^2 = 1 \times 1 = 1$
$1^3 = 1 \times 1 \times 1 = 1$

Result: 1 is equal to 1

The fraction $\frac{1}{2}$:

Squaring a positive fraction between 0 and 1 (for example, $\frac{1}{2}$) results in a smaller number.

Example: $(\frac{1}{2})^2 = \frac{1}{2} \times \frac{1}{2} = \frac{1}{4}$

Result: $\frac{1}{4}$ is smaller than $\frac{1}{2}$

Cubing a positive fraction between 0 and 1 (for example, $\frac{1}{2}$) also results in a smaller number.

Example: $(\frac{1}{2})^3 = \frac{1}{2} \times \frac{1}{2} \times \frac{1}{2} = \frac{1}{8}$

Result: $\frac{1}{8}$ is smaller than $\frac{1}{2}$

The number 0:

Squaring or cubing the number 0 is still zero.

Examples: $0^2 = 0 \times 0 = 0$
$0^3 = 0 \times 0 \times 0 = 0$

The fraction $-\frac{1}{2}$:

Squaring a negative fraction between 0 and –1 (for example, $-\frac{1}{2}$) results in a larger number.

Example: $(-\frac{1}{2})^2 = -\frac{1}{2} \times -\frac{1}{2} = \frac{1}{4}$

Result: $\frac{1}{4}$ is bigger than $-\frac{1}{2}$

This is no surprise because all negative numbers when squared get bigger. Why? Because they become positive and "jump to the other side" of zero on the number line.

Cubing a negative fraction between 0 and –1 (for example, $-\frac{1}{2}$) also results in a larger number.

Example: $(-\frac{1}{2})^3 = -\frac{1}{2} \times -\frac{1}{2} \times -\frac{1}{2} = -\frac{1}{8}$

Result: $-\frac{1}{8}$ is bigger than $-\frac{1}{2}$

Note that $-\frac{1}{8}$ is closer to zero than is $-\frac{1}{2}$. Thus, $-\frac{1}{8}$ is a larger number.

The number –1:

Squaring –1 results in a larger number.

Example: $(-1)^2 = -1 \times -1 = 1$

Result: 1 is bigger than –1

Cubing –1 results in the same number.

Example: $(-1)^3 = -1 \times -1 \times -1 = -1$

Result: –1 is equal to –1

The number –2:

Squaring –2 results in a larger number.

Example: $(-2)^2 = -2 \times -2 = 4$

Result: 4 is bigger than –2

Cubing –2 results in a smaller number.

Example: $(-2)^3 = -2 \times -2 \times -2 = -8$

Result: –8 is smaller than –2

117. Bigger or Smaller ( )

Which of the following is a true statement?

A) All positive numbers, when squared, get bigger.
B) All positive numbers, when cubed, get bigger.
C) Some negative numbers, when squared, stay the same.
D) All negative numbers, when cubed, get smaller.
E) All negative numbers, when squared, get bigger.

118. Plus-Zero ()

If $x > 0$, which of the following *could* be true?

I. $x^3 > x^2$

II. $x^2 = x$

III. $x^2 > x^3$

A) I only
B) I & II
C) II & III
D) All of the above
E) None of the above

119. Sub-Zero ()

If $x < 0$, which of the following *must* be true?

I. $x^2 > 0$

II. $x - 2x > 0$

III. $x^3 + x^2 < 0$

A) I only
B) I & II
C) II & III
D) All of the above
E) None of the above

Appendix V – Understanding How Variables Change

Variables change in one of four basic ways: they vary directly or indirectly and they vary proportionally or disproportionately. When variables vary directly, it means that when one goes up, the other goes up (and vice versa). When variables vary indirectly, it means that when one goes up, the other goes down (and vice versa). When one two or more variables vary proportionately (or linearly), it means that they vary by a "consistent" amount. For example, when one variable increases or decreases by a factor of 1 (doubles or is halved), another variable either increases or decreases by a factor of 1 (doubles or is halved).

When two or more variables vary disproportionately (non-linearly), it means that they vary by an "inconsistent" amount. For example, when one variable increases or decreases by a factor of 1, another variable either increases or decreases by a factor of more than 1. In this material, questions 1–10 present scenarios in which variables vary proportionately while questions 11–14 present scenarios in which variables vary disproportionately.

Let's take a very famous formula:

$$E = mc^2$$

It stands for: $Energy = mass \times (speed\ of\ light)^2$

We don't need to know what these variables mean in a technical sense to answer the following question.

Question: If the value of m is doubled, then what will happen to the value of E?

A) E will double.

B) E will quadruple.

C) E will stay the same.

D) E will be halved.

Well, if the value of m is doubled (that is, increased by a factor of 2), then E will also double (that is, be increased by a factor of 2). In other words, m and E vary directly. Note that c^2 is a constant and doesn't affect the outcome; we don't have to know the value of c.

120. Quiz on How Variables Change ( )

For questions 1–14, assume for simplicity's sake that A, b, and c are positive numbers greater than or equal to 1. Answers can be found on pages 270–272.

Formula 1: $A = \frac{b}{c}$

Questions 1–6 are based on Formula 1 above.

1. If b is doubled, then A will be

   A) doubled
   B) tripled
   C) quadrupled
   D) halved

2. If b is halved, then A will be

   A) doubled
   B) tripled
   C) quadrupled
   D) halved

3. If c is doubled, then A will be

   A) doubled
   B) tripled
   C) quadrupled
   D) halved

4. If c is halved, then A will be

   A) doubled
   B) tripled
   C) quadrupled
   D) halved

Formula 2: $A = \frac{1}{2}bc$

Questions 5–8 are based on Formula 2 above.

5. If b is doubled, then A will be

   A) doubled
   B) tripled
   C) quadrupled
   D) halved

6. If c is tripled, then A will be

   A) doubled
   B) tripled
   C) quadrupled
   D) halved

7. If c is halved, then A will be

   A) doubled
   B) tripled
   C) quadrupled
   D) halved

8. If b and c are both doubled, then A will be

   A) doubled
   B) tripled
   C) quadrupled
   D) halved

Formula 3: $A = \frac{1}{bc}$

Questions 9–10 are based on Formula 3 above.

9. If b is doubled, then A will be

   A) doubled
   B) tripled
   C) quadrupled
   D) halved

10. If c is halved, then A will be

   A) doubled
   B) tripled
   C) quadrupled
   D) halved

Formula 4: $A = \frac{b^2}{c}$

Questions 11–12 are based on Formula 4 above.

11. If b is doubled, then A will be

   A) doubled
   B) tripled
   C) quadrupled
   D) halved

12. If c is doubled, then A will be

A) doubled
B) tripled
C) quadrupled
D) halved

Formula 5: $A = \frac{b}{c^2}$

Questions 13–14 are based on Formula 5 above.

13. If b is doubled, then A will be

A) doubled
B) tripled
C) quadrupled
D) halved

14. If c is halved, then A will be

A) doubled
B) tripled
C) quadrupled
D) halved

Answers and Explanations to all 120 Problems

Chapter 1 – Basic Numeracy Ingredients

1. Antique Stamp ( )

Choice E. This is a "percentage of an original number" problem. The problem is essentially asking how many times larger, in percentage terms, is the current price compared with the original price.

$$\frac{\text{New}}{\text{Old}} \qquad \frac{\$150}{\$100} = 150\%$$

NOTE The calculation below is technically more accurate. The dollar signs (that is, $) cancel out and 1.5 must be multiplied by 100% in order to turn this decimal into a percentage and to reinstate the percentage sign in the final answer.

$$\frac{\$150}{\$100} = 1.5$$

$$1.5 \times 100\% = 150\%$$

2. Rise ( )

Choice D. This is a straight percentage increase. The formula for a percentage increase is as follows:

$$\frac{\text{New} - \text{Old}}{\text{Old}} \qquad \frac{\$150 - \$100}{\$100} = \frac{\$50}{\$100} = 50\%$$

3. Fall ( )

Choice D. This is a straight percentage decrease. The formula for a percentage decrease is as follows:

$$\frac{\text{Old} - \text{New}}{\text{Old}} \qquad \frac{\$100 - \$50}{\$100} = \frac{\$50}{\$100} = 50\%$$

4. Devalue ( )

Choice A. The formula for a percentage decrease to return to an original number is as follows:

$$\frac{\text{New} - \text{Old}}{\text{New}} \qquad \frac{\$125 - \$100}{\$125} = \frac{\$25}{\$125} = 20\%$$

5. Revalue ( )

Choice C. The formula for a percentage increase to return to an original number is as follows:

$$\frac{\text{Old} - \text{New}}{\text{New}} \qquad \frac{\$100 - \$75}{\$75} = \frac{\$25}{\$75} = 33\tfrac{1}{3}\%$$

6. Twist ( )

Choice D. There are two ways to approach this problem. The first is to pick a pair of simple numbers that make sense within the context of this problem. For example, say that the price of the stock was \$25 at the beginning of the year and \$100 at the end of the year; it therefore ended the year four times higher in price than it was at the start of the year.

The formula for a percentage increase is as follows:

$$\frac{\text{New} - \text{Old}}{\text{Old}} \qquad \frac{\$100 - \$25}{\$25} = \frac{\$75}{\$25} = 300\%$$

Alternatively, we can conceptualize three consecutive increases of 100% each. Since the stock price was originally 100% of its initial price, each of these three consecutive increases represents a 100% each or 300% in total. The stock ended the year 400% of what it was at the beginning of the year, but it increased 300% in price. The trap answer is thus 400%.

NOTE ⟜ How is the number 50 written as a percentage? Is it 0.005%, 0.5%, 50%, 500%, or 5000%? This is a rather tricky question that requires a solid understanding of how numbers are turned into percentages.

The answer is 5000%. How do we calculate this? One way is to take the original number 50 and add two zeros and slap on a percentage sign. Likewise, we can take the number 50 and move the decimal point two places to the right and add on a percentage sign. In order to visualize this process, start with something very familiar. For example, it's quite obvious that 0.5 is 50%. So how do we convert the decimal 0.5 to a percentage? We move the decimal two places to the right and add a percentage sign.

7. Microbrewery ()

Choice C. This problem is in essence asking about productivity: productivity = output ÷ hours. Assume that the original *output* and *hours* are each 100%. To calculate current output in percentage terms, add 70% to 100% to get 170%. To calculate current hours in percentage terms, subtract 20% from 100% to get 80%. Now divide 170% by 80% to get 212.5%. An even simpler calculation involves the use of decimals:

$$\frac{1.7}{0.8} = 2.125$$

$$2.125 \times 100\% = 212.5\%$$

To calculate the percentage increase:

$$\frac{\text{New} - \text{Old}}{\text{Old}} \qquad \frac{212.5\% - 100\%}{100\%} = \frac{112.5\%}{100\%} = 112.5\%$$

Encore! What if the wording to this problem had been identical except that the last sentence read:

> The year-end factory output per hour is what percentage of the factory output per hour at the beginning of the year?
>
> A) 50%
> B) 90%
> C) 112.5%
> D) 210%
> E) 212.5%

The answer would be choice E. This problem is not asking for a percentage increase, but rather "percentage of an original number."

Percentage of an original number:

$$\frac{\text{New}}{\text{Old}} \qquad \frac{212.5\%}{100\%} = 212.5\%$$

Again, the calculation below is more accurate. The percent signs (that is, %) cancel out and 2.125 must be multiplied by 100% in order to turn this decimal into a percentage and to reinstate the percentage sign in the final answer.

$$\frac{212.5\%}{100\%} = 2.125$$

$$2.125 \times 100\% = 212.5\%$$

NOTE ☞ The fact that two of the answer choices, namely choices C and E, are 100% apart alerts us to the likelihood that a distinction needs to be made between "percentage increase" and "percentage of an original number."

8. Gardener ( )

Choice B. View the area of the original rectangular garden as having each a width and length of 100%. The new rectangular garden has a length of 140% and a width of 80%. A 20% decrease in width translates to a width of 80% of the original.

Area of original garden:

$$\text{Area} = \text{length} \times \text{width} = 100\% \times 100\% = 100\%$$

Area of new garden:

$$\text{Area} = \text{length} \times \text{width} = 140\% \times 80\% = 112\%$$

Percentage change is as follows:

$$\frac{\text{New} - \text{Old}}{\text{Old}} \qquad \frac{112\% - 100\%}{100\%} = \frac{12\%}{100\%} = 12\%$$

Note: This shortcut calculation involves decimals:

$$1.4 \times 0.8 = 2.12$$
$$2.12 - 1.0 = 1.12$$
$$1.12 \times 100\% = 112\%$$

9. Train ( )

Choice C. In this problem, we must stick to the percentage increase formula. The correct calculation for the new speed is:

$$125\% \times 120\% = 150\%$$

$$\frac{\text{New} - \text{Old}}{\text{Old}} \qquad \frac{150\% - 100\%}{100\%} = \frac{50\%}{100\%} = 50\%$$

There are two trap answers. In choice B, we cannot just add 25% and 20% to get 45% because 20% of 125% is a bigger number than 20% of 100%. In choice E, 150% represents percentage of an original number, not percentage increase. It would have been the correct answer had the question instead asked: "The current speed is what percentage of the original speed?"

10. Broker ( )

Choice B. The trap answer is 20%. We cannot merely subtract these numbers (that is, 50% and 30%) because they are based on different wholes. The increase was based on the original 100% figure. The decrease was based on the 150% figure.

The actual answer is obtained by multiplying 150% by 70% and subtracting 100% from this total. That is: 150% × 70% = 105%; 105% – 100% = 5%.

NOTE If this particular problem were to appear on an actual math test and you needed to guess at an answer, how could you guess strategically? In other words, what are the classic guessing strategies for use on problems presented in a multiple choice format? First, numerical answers to multiple-choice problems are typically presented in ascending or descending order in terms of size. If you must guess, the key strategies of elimination include: (1) eliminate an answer that looks different from the others, (2) eliminate answers which look too big or too small, that is, extreme answers, and (3) eliminate answers which contain the same or similar numbers as given in the question or are easy derivatives of the numbers used in the problem. By easy derivatives, think in terms of addition and subtraction, not multiplication and division. For example, in the problem at hand, eliminate –5% because it is negative, and thus different from the other positive numbers. Eliminate 80% because it is much bigger than any other number (extreme). Eliminate 20% because it is an easy derivative of the numbers mentioned in the question, (that is, 50% less 30%). You would then guess choices B or C.

Most importantly, remember that these classic guessing strategies are guidelines, not infallible yardsticks.

11. Double Discount (🌶)

Choice D.

80% × 90% = 72%
100% – 72% = 28%

A 20% discount followed by a 10% discount amounts to a 28% discount. We cannot simply add 20% to 10% to get 30%. The rationale for this is encapsulated by Tip #2—You can't add (or subtract) the percentages of different wholes. The first whole is 100% and the second whole is 80%.

12. Net Effect #1 (🌶)

Choice C.

90% × 110% = 99%
100% – 99% = 1% net decrease

Say that the original price was 100%. It decreases from 100% to 90%. When it increases by 10%, it is now 99% of the original price. This works out simply because 10% of 100% is a greater percentage than 10% of 90%. Note that using decimals rather than percentages makes the calculation even simpler.

0.9 × 1.1 = 0.99
1.0 – 0.99 = 0.01
0.01 × 100% = 1% net decrease

13. Net Effect #2 ( )

Choice C.

$$120\% \times 80\% = 96\%$$
$$100\% - 96\% = 4\% \text{ net decrease}$$

Say that the original price was 100%. It increases from 100% to 120%. When it decreases by 20%, it is now 96% of the original price. This works out simply because 20% of 120% is a greater percentage than 20% of 100%. Again, making the calculation using decimals rather than percents is simpler.

$$1.2 \times 0.8 = 0.96$$
$$1.0 - 0.96 = 0.04$$
$$0.04 \times 100\% = 4\% \text{ net decrease}$$

14. Silver ( )

Choice C. Let's use 100% as a base. A 100% increase gives us 200%. A 50% decrease from 200% gives us 100%—the original base. The reason things work out this way is that the 100% increase is based on a smaller base (that is, 100%) as compared with the 50% decrease, which is based on a larger base (that is, 200%).

Quick calculation: $2.00 \times 0.5 = 100$

15. Growth ( )

Choice A. In fact, it will take approximately 7 years.

Calculation for \$1@10% for 7 years:

$$\$1 \times (1.10)^7 = \$1 \times 1.95 \cong \$2$$

Calculation for \$1@7% for 10 years:

$$\$1 \times (1.07)^{10} = \$1 \times 1.96 \cong \$2$$

The "Seven-Ten" Rule of Finance is so called because a business growing at 10% a year will double in size in 7 years. Obviously this is the result of compounded growth. Likewise, a business growing at 7% a year will double in size in approximately 10 years. The trap answer is 10 years—10 years at 10% per year equals 100%. But because the company is growing at 10% per year, the growth is compounding, and the 10% figure is being calculated on a different whole (that is, growth plus base) for each succeeding year.

16. Medical School ( )

Choice C. Percents and numbers do not always mix. The seventy-five percent of male students that are actually enrolled in medical school cannot be compared with half of all female applicants. This question only makes sense if we know the percentage of females accepted and compare this figure with the percentage of males accepted. A likely scenario is that 75% of all students in medical school are male because 75% of all medical school applicants are male; 25% of all students accepted to medical school are female because 25% of the total applicants are female—little surprise.

Take this hypothetical example. Say that 10,000 applicants apply for a spot at a top medical school. What if 25% of the applicant pool is female while 75% is male? This means that 2,500 females and 7,500 males will apply for admission.

Let's assume that the acceptance rate is 5% (only 1 in 20 applicants is accepted). The opening class will consist of 500 students (that is, $10{,}000 \times 5\% = 500$). Common sense tells us that with rules of fair selection in place, the opening class should be 75% male and 25% female, which directly reflects the applicant pool. Thus, the number of females in the opening class should be 125 ($500 \times 25\%$) while the number of males in the opening class should be 375 ($500 \times 75\%$).

Now, relate hypothetical figures within the context of the original passage. "The facts in the report speak for themselves: Seventy-five percent of all students in medical school are male, but fewer than half of all the 2,500 female applicants reach their goal of being admitted to medical school." Of course less than half of all females are admitted; only 1 in 20 students overall is accepted! Choices A, B, D, and E are all irrelevant to the issue at hand. In choice B, there is no reason to suspect that the yield rates (percentage of offers given by medical schools that are accepted by students) would in any significant way be affected by gender. So although it might warrant some consideration, it is by no means a close contender with choice C.

17. White-Collar Crime ( )

Choice E. The number of white-collar crimes last year is not the same as the rate of increase in white-collar crime last year. Say, for example, that there were 900 cases of white-collar crime the year before last and there were 1,000 cases of white-collar crime this past year. This means that 50% of the increase, or 50 cases, is due to unemployment, often aggravated by alcoholism.

In terms of the number of white-collar crimes that occurred last year, most of them occurred due to desperation that results from overwork. This means that more than half of the 1,000 cases last year, or 500-plus cases, were due to loneliness that results from overwork.

The author's criticism of Dr. Noah is not valid because we cannot compare the 50 cases representing the percentage of rate increase with the 500 cases representing white-collar crime due to the desperation that results from overwork. In other words, the majority of white-collar crime is still due to the desperation that results from overwork even though the greatest increase (trend) has been caused by unemployment aggravated by alcoholism.

18. Fly Fishing ( )

Choice E. Percentages are really fractions. Fractions are composed of a numerator (the top number in a fraction) and denominator (the bottom number in a fraction). A percentage is increased either by increasing the numerator of that fraction or by decreasing a denominator of that fraction, or doing both simultaneously. Choices A, B, C, and D could be true. However, choice E could not be true. Choice E must lead to a percentage decrease, not a percentage increase.

A) $$\frac{\uparrow \text{number of people fly fishing}}{\downarrow \text{sport-fishing population}} = \uparrow \%$$

B) $$\frac{\uparrow \text{number of people fly fishing}}{\text{sport-fishing population (remains the same)}} = \uparrow \%$$

C) $$\frac{\uparrow \text{number of people fly fishing}}{\uparrow \text{sport-fishing population}} = \uparrow \%$$

Could be true as long as the numerator increases incrementally faster than the denominator.

D) $$\frac{\downarrow \text{number of people fly fishing}}{\downarrow \text{sport-fishing population}} = \uparrow \%$$

Could be true as long as the denominator decreases incrementally faster than the numerator.

E) $$\frac{\text{fly fishing population (remains the same)}}{\uparrow \text{sport-fishing population}} = \downarrow \%$$

19. Lottery ( )

Choice C. We can't tell. It all depends on how many people are in each income category. For example, since we expect fewer people to be in the over \$100,000 income category (compared with the other income categories), we also expect fewer ticket purchases by individuals in this category. This makes it impossible to know the ratio or percentage of people in each category who buy lottery tickets. In order to make an actual, meaningful comparison, we would need to divide the number of lottery tickets bought by individuals in each income category by the number of individuals in each category.

20. Iraq ( )

Choice D. Obviously the chances of an American solider being killed serving in Iraq during his or her first three-year period of the Second Gulf War were greater than the chances of a U.S. civilian being killed driving on the roads of America during the same three-year period. We cannot directly compare raw numbers. A more realistic comparison would be to compare combat deaths to the total number of combat soldiers while comparing traffic deaths to the total number of U.S. drivers. The following calculations show the role that each denominator plays in each emerging ratio:

$$\frac{\text{Military Combat Deaths}}{\text{Total Number of Active Soldiers in Iraq}} = \frac{2{,}300}{130{,}000} = 1.7\%$$

$$\frac{\text{Auto Traffic Deaths}}{\text{Total Number of US drivers}} = \frac{100{,}000}{50{,}000{,}000} = 0.2\%$$

From these statistics we see that it was 8.5 times more dangerous serving as a combatant in Iraq than it was to be a driver in America during the same three-year period.

$$\frac{\text{Death Rate (Military Combatants in Iraq)}}{\text{Death Rate (US Drivers)}} = \frac{1.7}{0.2} = 8.5$$

The scope of the two situations being compared is deaths serving in Iraq and deaths driving in America during a three-year period. The prospect of gun-related deaths (choice A) is essentially out of our argument's scope. And as deaths are unfortunately deaths, it is essentially irrelevant to consider "driving deaths resulting from high speeds" (choice B), "deaths caused by accidents" (choice C), or "percentage of time driving a car or serving in the armed forces" (choice E). This last choice is especially mind warping. Of course, it doesn't matter whether one drives an hour one day or serves a full day in Iraq. The argument is objectively comparing the death rate in one situation to the death rate in another. Besides, the fact that the average person drives for only a small portion of each day seems to help corroborate the original argument, not help refute it.

21. Military Expenditures (🌶🌶)

Choice B. This problem highlights the concept of the growing pie. The apparent inconsistency is resolved if it is true that the portion or percentage of the national budget represented by military expenditures has increased by a greater amount than actual expenditures. If, for example, the national budget has increased tenfold during the past two decades and the amount of actual expenditures (inflation adjusted) has increased by 1 percent per year, then the portion devoted to military expenditure has likely shrunk. Even if actual dollars spent on military expenditures are increasing each year, as long as overall (or total) expenditures are increasing more, the amount of military expenditures, in percentage terms, is falling.

The opposite of a growing pie is a shrinking pie. If this problem were a shrinking pie, actual dollars spent on military expenditures would be decreasing each year, but as long as overall expenditures were shrinking faster, the amount of military expenditures, in percentage terms, would be rising.

22. Fiction Books (🌶🌶🌶)

Choice E. This problem illustrates the ultimate tangling of numbers with percentages. Choice A is out of scope because we don't know about people with the highest levels of education; we only know about members of households with higher levels of education (that is, HEL—high education level households) and members of households with lower levels of education (that is, LEL—low education level households). Also, we do not know whether people really buy, are given, or inherit these books. Choices B, C, and D are out because we do not know for sure whether HEL households or LEL households have more or less fiction versus nonfiction books. We also do not know whether members of HEL households have more or fewer nonfiction books as compared with members of LEL households. The only thing we do know for sure is that if members of HEL households have more books in total than members of LEL households and a higher percentage of fiction versus nonfiction, then it must be true that HEL households contain a greater number of fiction books than LEL households.

Each of the three scenarios that follow illustrates a significant possibility based on hypothetical numbers of books and percentages. Again, upon examination, the only thing that must be true is that for any given scenario the number of fiction books in HEL households must be greater than the number of fiction books in LEL households. Let's use 100 books and 50 books to test things out. The number of fiction books for HEL households, column (1), will always be greater than the number of fiction books for LEL households, column (3).

Scenario #1: HEL households have 80% fiction books and 20% nonfiction books while LEL households have 50% fiction books and 50% nonfiction books.

HEL Households 100 books		LEL Households 50 books	
(1) (Fiction) 80 books	(2) (Nonfiction) 20 books	(3) (Fiction) 25 books	(4) (Nonfiction) 25 books

Scenario #2: HEL households have 70% fiction books and 30% nonfiction books while LEL households have 50% fiction books and 50% nonfiction books.

HEL Households 100 books		LEL Households 50 books	
(1) (Fiction) 70 books	(2) (Nonfiction) 30 books	(3) (Fiction) 25 books	(4) (Nonfiction) 25 books

Scenario #3: Get ready for this! HEL households could have 40% fiction books and 60% nonfiction books while LEL households have 20% fiction books and 80% non-fiction books. This is possible because the percentage of fiction books for HEL households is still greater—40%—than the percentage of fiction books for LEL households—20%.

HEL Households 100 books		LEL Households 50 books	
(1) (Fiction) 40 books	(2) (Nonfiction) 60 books	(3) (Fiction) 10 books	(4) (Nonfiction) 40 books

23. WAA ( )

Choice A. We need the statement that would not provide a logical objection to the WAA's conclusion. Although it's probably true that countries with more miles of road are likely to have more traffic deaths, the ranking of deaths is per mile driven. This ratio makes it possible to compare various countries irrespective of how many miles of roadway they have.

The wrong answer choices are all valid objections because they point to reasons other than driving ability for the number of deaths per mile driven. The difference in road safety from one country to another (choice B) is a factor in accidents. How well a car is maintained (choice C) can also be a factor in whether a driver has an accident. A difference in terrain (choice D) means good drivers in mountainous countries may be more likely to have a fatal accident. And a higher number of passengers per car (choice E) might lead to a high number of traffic fatalities, even where the frequency of accidents is relatively low.

24. Lemonade ( )

Choice E. The ratio cannot be determined. The trap answer is choice A. We cannot simply add the mixtures, as would be the case if the mixtures were of equal size (that is, 1:2 + 2:3 = 3:5). If the original mixtures vary considerably in size, the final ratio will also vary considerably.

25. Eggs ( )

Choice D. As it is impossible to have 2.4 broken eggs (or 9.6 unbroken eggs), choice D is correct.

A) 1:1 12 eggs = 6 broken $(\frac{1}{2} \times 12)$; 6 unbroken $(\frac{1}{2} \times 12)$

B) 1:2 12 eggs = 4 broken $(\frac{1}{3} \times 12)$; 8 unbroken $(\frac{2}{3} \times 12)$

C) 1:3 12 eggs = 3 broken $(\frac{1}{4} \times 12)$; 9 unbroken $(\frac{3}{4} \times 12)$

D) 1:4 12 eggs = 2.4 broken $(\frac{1}{5} \times 12)$; 9.6 unbroken $(\frac{4}{5} \times 12)$

E) 1:5 12 eggs = 2 broken $(\frac{1}{6} \times 12)$; 10 unbroken $(\frac{5}{6} \times 12)$

26. Elk Herd ( )

Choice B. Since the ratio of male elk to female elk is 16 to 9, it is true that 16 of every 25 (that is, 16 + 9) elk are male. In other words, a part-to-part ratio of 16 to 9 becomes a part-to-whole ratio of 16 to 25. A ratio of 16 : 25 equals 64 percent (that is, $\frac{16}{25} = 64\%$).

27. Rum & Coke ( )

Choice B. The trap answer is choice A because it erroneously adds component parts of the two different ratios. That is: 1 : 2 + 1 : 3 does not equal 2 : 5. This could only be correct if ratios represent identical volumes. We cannot simply add two ratios together unless we know the numbers behind the ratios.

	Total	Rum	Coke
Solution #1	6	2	4
Solution #2	32	8	24
Totals		10	28
		Simplify☺	
Final Ratio		5	14

The ratio of 10 : 28 simplifies to 5 : 14.

Supporting calculations:

$6 \times \frac{1}{3} = 2$ Two ounces of rum in solution #1.

$6 \times \frac{2}{3} = 4$ Four ounces of Coke in solution #1.

$32 \times \frac{1}{4} = 8$ Eight ounces of rum in solution #2.

$32 \times \frac{3}{4} = 24$ Twenty-four ounces of Coke in solution #2.

28. Toledo Prep ( )

Choice C. Likely the best way to approach this problem is to pick numbers. A faculty-to-student ratio of 1 to 7 means that for every 80 persons, 10 are faculty members and 70 are students:

Breakdown: $80 \times \frac{1}{8} = 10$ faculty members; $80 \times \frac{7}{8} = 70$ students

Keeping with these same numbers, if 10 people are faculty members, then 2 are female and 8 are male: $10 \times \frac{1}{5} = 2$ females and $10 \times \frac{4}{5} = 8$ males. For every 70 students, 30 are female and 40 are male:

Breakdown: $70 \times \frac{3}{7} = 30$ females; $70 \times \frac{4}{7} = 40$ males

Therefore: $\frac{\text{females}}{\text{total persons}} = \frac{2+30}{10+70} = \frac{32}{80} = \frac{4}{10} = 40\%$

Another way to organize the information in this problem is with a matrix. Again, assume that the total number of students and faculty is 80.

	Faculty	Students	
Male	8	40	48
Female	2	30	32
	10	70	80

29. Deluxe ( )

Choice D. This is a more complex problem than *Toledo Prep* (Problem 28), but conceptually similar. We know that there are 24 liters of fuchsia paint in a ratio of 5 parts red to 3 parts blue. The first step then is to break the 24 liters of fuchsia into actual amounts (liters) of "red" and "blue" paint. This requires using a part-to-whole ratio (that is, three-eighths blue and five-eighths red; $\frac{3}{8}$ and $\frac{5}{8}$ respectively).

Blue: 5 parts red to 3 parts blue:

$$\frac{3}{5+3} = \frac{3}{8} \quad \rightarrow \quad \frac{3}{8} \times 24 = 9 \text{ liters (of blue paint)}$$

Red: 5 parts red to 3 parts blue:

$$\frac{5}{5+3} = \frac{5}{8} \quad \rightarrow \quad \frac{5}{8} \times 24 = 15 \text{ liters (of red paint)}$$

Our final ratio is a part-to-part ratio, comparing liters of red and blue paint in fuchsia to the ratio of red and blue paint in mauve. So the final formula, expressed as a proportion, becomes:

$$\frac{15_{\text{red}}}{9_{\text{blue}} + x_{\text{blue}}} = \frac{3_{\text{red}}}{5_{\text{blue}}}$$

$5(15) = 3(9 + x)$

$75 = 3(9 + x)$

$75 = 27 + 3x$

$75 - 27 = 3x$

$48 = 3x$

$3x = 48$

$x = 16$ liters (of blue paint)

NOTE ☞ In the above equation, we simply reverse the equation and "$48 = 3x$" becomes "$3x = 48$." The rationale behind this practice may be viewed in either of two ways. The first is based in logic. If A equals B, then it is true that B equals A. The second is based in algebra and unfolds as follows:

$48 = 3x$

$-3x = -48$ [switching sides changes the signs]

$(-1)(-3x) = (-1)(-48)$ [multiply both sides by -1]

$3x = 48$

In short, when solving for a single variable such as x, our practical goal is to get x on one side of the equation and all other terms on the opposite side of the equation. Typically, this involves isolating x on the left-hand side of the equation while placing all the other terms on the right-hand side of the equation. There are two useful concepts to keep in mind when manipulating elements of a formula in order to achieve this objective. The first is that bringing any number (or term) across the equals sign, changes the sign of that number (or term). That is, positive numbers become negative and negative numbers become positive. The second useful concept is that whatever we do to one term in an equation, we must do to every other term in that same equation (in order to maintain the equivalent value of all terms in the equation). So in the above equation, we may choose to multiply each side of the equation by -1 in order to cancel the negative signs. We do this because we always want to solve for a positive x value.

30. Exchange ( )

Choice D. Since one Philippine peso equals 0.15 Hong Kong dollars, one Hong Kong dollar divided by 0.15 equals 6.67 pesos.

Ratio Structure:

$$\frac{A}{B} :: \frac{C}{D}$$

$$\frac{\text{Part}}{\text{Part}} :: \frac{\text{Whole}}{\text{Whole}}$$

Applied to this problem:

$$\frac{\text{Philippine Peso}}{\text{Hong Kong Dollar}} :: \frac{\text{Philippine Peso}}{\text{Hong Kong Dollar}}$$

Calculation:

$$\frac{1_{\text{PP}}}{0.15_{\text{HK\$}}} = \frac{x_{\text{PP}}}{1_{\text{HK\$}}}$$

$$1(1) = 0.15(x)$$

$$1 = 0.15x$$

$$0.15x = 1$$

$$x = \frac{1}{0.15}$$

$$x = 6.67 \text{ pesos}$$

31. Sale Time ()

Choice D. The quick method is to divide \$12 by 30%. That is, $\$12 \div 0.30 = \$40$. The ratio and proportion method follows below:

$$\frac{\$12}{30\%} = \frac{x}{100\%}$$

$$100\%(\$12) = 30\%(x)$$

$$30\%(x) = 100\%(\$12)$$

$$\frac{1}{\cancel{30\%}} \times \cancel{30\%}(x) = \frac{1}{30\%} \times 100\%(\$12)$$

$$x = \frac{\$12(100\%)}{30\%}$$

$$x = \frac{\$12(1.0)}{0.3}$$

$$x = \$40$$

Note, per the penultimate step above, we may choose to convert percentages (that is, 100% and 30%) to decimals (that is, 1.0 and 0.3) for ease of computation.

Extra: Suppose a problem asked: "A man bought a pair of shoes for \$28 during a 30% off sale. What was the retail price of the shoes?"

Again, the quick method is to divide \$28 by 70%. That is, \$28 ÷ 0.70 = \$40. This is the same problem but viewed from a different angle. Here's the more formal ratio-and-proportion method:

$$\frac{\$28}{70\%} = \frac{x}{100\%}$$

$$100\%(\$28) = 70\%(x)$$

$$70\%(x) = 100\%(\$28)$$

$$\frac{1}{\cancel{70\%}} \times \cancel{70\%}(x) = \frac{1}{70\%} \times 100\%(\$28)$$

$$x = \frac{\$28(100\%)}{70\%}$$

$$x = \frac{\$28(1.0)}{0.7}$$

$$x = \$40$$

32. Landsdown ( )

Choice B. The shortcut method is to merely divide 18,000 by 12% or 0.12. If this is not obvious, we must rely on setting up a classic proportion. The final calculation will be same in either case.

$$\frac{12\%}{100\%} = \frac{18,000}{x}$$

$$12\%x = 100\%(18,000)$$

$$\frac{1}{\cancel{12\%}} \times \cancel{12\%}(x) = \frac{1}{12\%} \times 100\%(18,000)$$

$$x = \frac{18,000(100\%)}{12\%}$$

$$x = \frac{18,000(1.0)}{0.12}$$

$$x = 150,000$$

33. Shrinkage ( )

Choice C. Arguably the fastest way to solve this problem is to set it up as a ratio or proportion.

$$\frac{75\%}{100\%} = \frac{24}{x}$$

$$75\%(x) = 100\%(24)$$

$$\frac{1}{\cancel{75\%}} \times \cancel{75\%}(x) = \frac{1}{75\%} \times 100\%(24)$$

$$x = \frac{24(100\%)}{75\%}$$

$$x = \frac{24(1.0)}{0.75}$$

$$x = 32 \text{ ounces}$$

34. Product A ( )

Choice D.

$$\frac{120\%}{100\%} = \frac{x}{\$40}$$

$$120\%(\$40) = 100\%(x)$$

$$100\%(x) = 120\%(\$40)$$

$$\frac{1}{\cancel{100\%}} \times \cancel{100\%}(x) = \frac{1}{100\%} \times 120\%(\$40)$$

$$x = \frac{\$40(120\%)}{100\%}$$

$$x = \frac{\$40(1.2)}{1.0}$$

$$x = \$48$$

Note that we cannot divide \$40 by 0.8 to get \$50. This is incorrect because the reciprocal of 1.2 is 0.83, not 0.8. As illustrated by Tip #11, $\frac{120\%}{100\%}$ is not the same as $\frac{100\%}{80\%}$.

35. Product B ( )

Choice C.

$$\frac{120\%}{100\%} = \frac{\$50}{x}$$

$$120\%(x) = 100\%(\$50)$$

$$\frac{1}{\cancel{120\%}} \times \cancel{120\%}(x) = \frac{1}{120\%} \times 100\%(\$50)$$

$$x = \frac{\$50(100\%)}{120\%}$$

$$x = \frac{\$50(1.0)}{1.2}$$

$$x = \$41.67$$

36. Product C ( )

Choice E.

$$\frac{100\%}{80\%} = \frac{x}{\$40}$$

$$100\%(\$40) = 80\%(x)$$

$$80\%(x) = 100\%(\$40)$$

$$\frac{1}{\cancel{80\%}} \times \cancel{80\%}(x) = \frac{1}{80\%} \times 100\%(\$40)$$

$$x = \frac{\$40(100\%)}{80\%}$$

$$x = \frac{\$40(1.0)}{0.8}$$

$$x = \$50$$

37. Product D ( )

Choice B.

$$\frac{100\%}{80\%} = \frac{\$50}{x}$$

$$100\%(x) = 80\%(\$50)$$

$$\frac{1}{\cancel{100\%}} \times \cancel{100\%}(x) = \frac{1}{100\%} \times 80\%(\$50)$$

$$x = \frac{\$50(80\%)}{100\%}$$

$$x = \frac{\$50(0.8)}{1.0}$$

$$x = \$40$$

38. Diners ( )

Choice B. This nice, round number represents the cost before tax and tip.

Calculation of the cost <u>before</u> tip:

$$\frac{x}{\$264} = \frac{100\%}{120\%}$$

$$120\%(x) = 100\%(\$264)$$

$$\frac{1}{\cancel{120\%}} \times \cancel{120\%}(x) = \frac{1}{120\%} \times 100\%(\$264)$$

$$x = \frac{\$264(100\%)}{120\%}$$

$$x = \frac{\$264(1.0)}{1.2}$$

$$x = \$220$$

Calculation of the cost <u>before</u> tip <u>and</u> taxes:

$$\frac{x}{\$220} = \frac{100\%}{110\%}$$

$$110\%(x) = 100\%(\$220)$$

$$\frac{1}{\cancel{110\%}} \times \cancel{110\%}(x) = \frac{1}{110\%} \times 100\%(\$220)$$

$$x = \frac{\$220(100\%)}{110\%}$$

$$x = \frac{\$220(1.0)}{1.1}$$

$$x = \$200$$

NOTE The quick method is to divide \$264 by 1.2 and then by 1.1. That is, (\$264 ÷ 1.2) ÷ 1.1 = \$200. Likewise, we can divide \$264 by 1.32. That is, \$264 ÷ (1.2 × 1.1) = \$264 ÷ 1.32 = \$200.

39. Investments ()

Choice E. Below are the calculations for gain and loss expressed as mathematical proportions.

Gain on sale of property A:

$$\frac{120\%}{100\%} = \frac{\$24{,}000}{x}$$

$$120\%(x) = 100\%(\$24{,}000)$$

$$\frac{1}{\cancel{120\%}} \times \cancel{120\%}(x) = \frac{1}{120\%} \times 100\%(\$24{,}000)$$

$$x = \frac{\$24{,}000(100\%)}{120\%}$$

$$x = \frac{\$24{,}000(1.0)}{1.2}$$

$$x = \$20{,}000$$

Gain: \$24,000 – \$20,000 = \$4,000

This gain represents the sales price less original purchase price.

Loss on the sale of property B:

$$\frac{100\%}{80\%} = \frac{x}{\$24{,}000}$$

$$100\%(\$24{,}000) = 80\%(x)$$

$$80\%(x) = 100\%(\$24{,}000)$$

$$\frac{1}{\cancel{80\%}} \times \cancel{80\%}(x) = \frac{1}{80\%} \times 100\%(\$24,000)$$

$$x = \frac{\$24,000(100\%)}{80\%}$$

$$x = \frac{\$24,000(1.0)}{0.8}$$

$x = \$30,000$

Loss: \$30,000 – \$24,000 = \$6,000

This loss represents the original purchase price less the amount received from the sale.

Therefore, we have an overall loss of \$2,000 (net \$6,000 loss and \$4,000 gain). Note that the following provides shortcut calculations:

Calculation of gain:

\$24,000 ÷ 1.2 = \$20,000
Gain: \$24,000 – \$20,000 = \$4,000

Calculation of loss:

\$24,000 ÷ 0.8 = \$30,000
Loss: \$30,000 – \$24,000 = \$6,000

Again, we have an overall loss of \$2,000 (net \$6,000 loss and \$4,000 gain).

40. Subtle ( )

Choice D. Dividing 100 by 0.75 is the same as multiplying 100 by the reciprocal of 0.75. The reciprocal of 0.75 is 1.33, not 1.25! This problem highlights the importance of Tip #12 (see *Chapter 1*).

41. Topsy-Turvy ( )

Choice C. Multiplying a number by 1.2 is the same as dividing that same number by the reciprocal of 1.2, that is, 0.83. Note how answer choice B offers a tempting but incorrect answer of 0.80.

42. Partners ( )

Choice E. The associative law applies to multiplication and addition; it states that "regroupings" don't affect the outcome of a mathematical expression.

Incorrect answer choices A, B, C, and D all highlight the associative property of multiplication—they are equivalent expressions for $a \times (b \times c \times d)$. Correct answer choice E is not an equivalent expression of $a \times (b \times c \times d)$.

43. Bargain ( )

Choice C. This example highlights the commutative property of multiplication. It doesn't matter in which order we multiply numbers, for the answer remains the same. In this case, a 10% discount followed by a 30% discount is the same as selling a given product for 90% of its original price multiplied by 70% of its original price (90% × 70% = 63%). A 30% discount followed by a 10% discount is also the same as selling a given product for 70% of its original price times 90% of its original price (70% × 90% = 63%).

Contrast this problem with *Double Discount* (Problem 11). *Bargain* does not require the actual amount of overall discount, but rather the relationship between the two discounts. For the record, a 10% discount followed by a 30% discount is the same as one 37% discount (not 40% as offered by trap answer choice B).

44. Inflation ( )

Choice C. This problem also highlights the commutative property of multiplication in which *order* doesn't matter.

$$120\% \times 110\% = 132\%$$
$$110\% \times 120\% = 132\%$$

It does not matter the *order* in which we multiply numbers, the answer remains the same. In this case, a 20% inflationary increase followed by a 10% inflationary increase is the same as an inflationary increase of 10% followed by an inflationary increase of 20%. Either way we have an overall inflationary increase of 32%.

Again, this problem does not require the actual amount of overall increase but rather the relationship between the two inflationary increases. For the record, a 10% discount followed by a 30% discount is the same as a 30% discount followed by a 10% discount. Test: $0.9 \times 0.7 = 0.63 \times 100\% = 63\%$ and $0.7 \times 0.9 = 0.63 \times 100\% = 63\%$. Either way, we have an overall discount of 37%.

Chapter 2 – Wonderful Math Recipes

45. Country Club ( )

Choice A.

Two-Groups Formula		Solve
Add:	Group A (Tennis)	700
Add:	Group B (Golf)	500
Less:	Both	<?>
Add:	Neither	1,500
	Total	2,500

Calculation:

Group A + Group B – Both + Neither = Total
$700 + 500 - x + 1{,}500 = 2{,}500$
$2{,}700 - x = 2{,}500$
$x = 200$

That is, 200 people play both tennis and golf. Below is the visual or Venn-diagram representation for this problem.

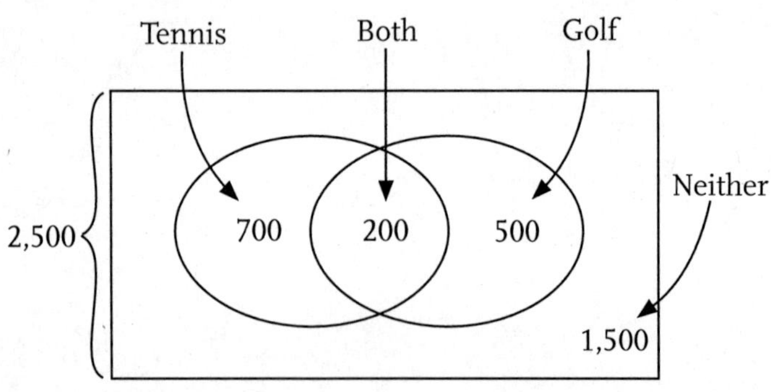

46. Nova vs. Rebound ()

Choice D. We cannot assume that a person liked either Nova or Rebound running shoes without taking into account the possibility that a person could like both brand name shoes at the same time. This situation results in overlap and accounts for the fact that the percentage appears to be greater than 100% (that is, 53% + 47% + 24% = 124%).

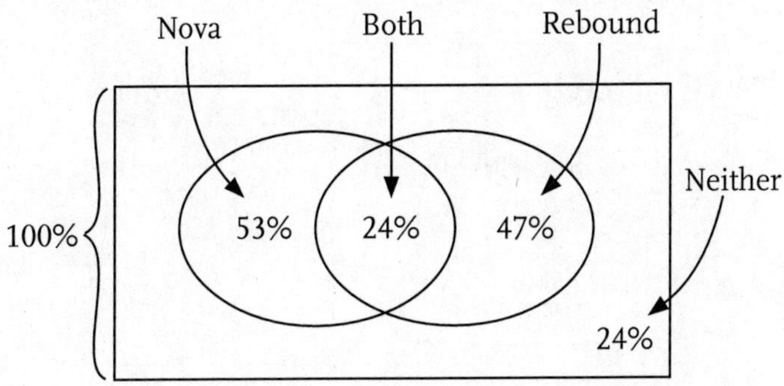

The following is the mathematical solution to this problem.

Two-Groups Formula		Solve
Add:	Group A (Nova)	53%
Add:	Group B (Rebound)	47%
Less:	Both	$<x>$
Add:	Neither	24%
	Total	100%

Calculation:

$$\text{Group A} + \text{Group B} - \text{Both} + \text{Neither} = \text{Total}$$

$$53\% + 47\% - x + 24\% = 100\%$$

$$124\% - x = 100\%$$

$$x = 24\%$$

Therefore, the number of people liking both Nova and Rebound running shoes is 24%. This problem is equally strange and tricky because 24% turns out to be the identical figure for the "both" and "neither" components in the group formula. Although this is indeed a peculiar result, it is certainly a mathematical possibility.

47. Science ( )

Choice B. Six people are enrolled in both chemistry and physics.

Two-Groups Formula		**Solve**
Add:	Group A (Chemistry)	22
Add:	Group B (Physics)	19
Less:	Both	<*x*>
Add:	Neither	0
	Total	35

Calculation:

Group A + Group B – Both + Neither = Total
$22 + 19 - x + 0 = 35$
$41 - x = 35$
$x = 6$

Below is an alternative algebraic calculation. This alternative algebraic approach works because everyone is enrolled in either chemistry or physics or both, and there is no possibility of someone not being enrolled in either. Here the variable *b* represents those students enrolled in both chemistry and physics.

$(22 - b) + b + (19 - b) = 35$
$41 - b = 35$
$b = 6$

48. Standardized Test ()

Choice A.

Two-Groups Formula		Solve
Add:	Group A (1st Question)	85%
Add:	Group B (2nd Question)	75%
Less:	Both	<65%>
Add:	Neither	x%
	Total	100%

Calculation:

Group A + Group B – Both + Neither = Total
$85\% + 75\% - 65\% + x = 100\%$
$95\% + x = 100\%$
$x = 5\%$

Five percent of the test takers got neither the first question nor the second question correct. In the Venn-diagram pictorial below, note that 100 percent is what is in the "box"; it includes Q1 and Q2 and "neither," but it must not include "both" because "both" represents overlap that must be subtracted out; otherwise the overlap will be double counted. In other words, either Q1 may include the 65% as part of its 85% or Q2 may include the 65% as part of its 75%, but Q1 and Q2 cannot both claim it. Another way to view this problem is to break it up analytically as follows: The number of students who only got Q1 correct is 20% (85% – 65%) while the number of students who only got Q2 correct is 10% (75% – 65%). Since 65% of students got both questions correct, the number of students who got one or the other question correct is: 20% + 10% + 65% = 95%; 5% of students got neither question correct.

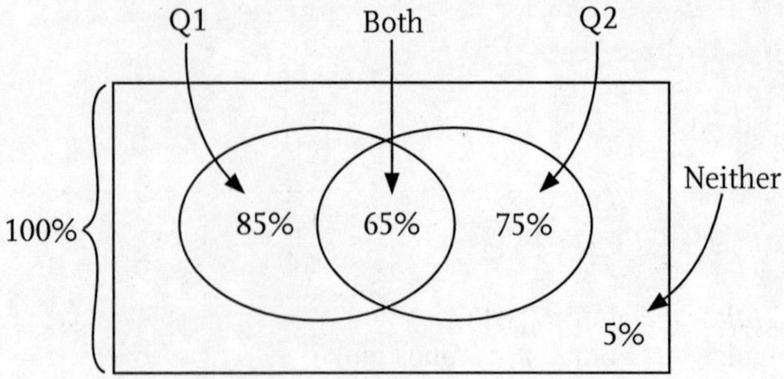

49. Language Classes (🌶🌶🌶)

Choice D. This problem is more difficult and merits a three-chili rating insofar as it requires an answer that is expressed in terms of an algebraic expression.

Use the classic "two-groups" formula:

Group A + Group B – Both + Neither = Total

Applied to the problem at hand:

Spanish + French – Both + Neither = Total Students

$$S + F - B + N = X$$

$$N = X + B - S - F$$

Therefore, expressed as a percent:

$$\text{Neither} = 100 \times \frac{X + B - S - F}{X}$$

50. Valley High ()

Choice B.

Three-Groups Formula		**Solve**
Add:	Biology	4
Add:	Chemistry	4
Add:	Physics	4
Less:	Biology & Chemistry	<2>
Less:	Biology & Physics	<2>
Less:	Chemistry & Physics	<2>
Add:	Biology & Chemistry & Physics	1
Add:	None of Biology, Chemistry or Physics	0
	Total	x

Calculation:

$$B + C + P - BC - BP - CP + BCP + \text{None} = \text{Total}$$
$$4 + 4 + 4 - 2 - 2 - 2 + 1 + 0 = x$$
$$x = 7$$

Seven students are involved in independent study projects. The following Venn-diagram confirms that there are seven students, as indicated by letters standing for names of students. According to our calculation above, student G (who is taking all of Biology, Chemistry, and Physics) needs to be added back because he or she has been subtracted out one too many times as a result of deducting the three double overlaps.

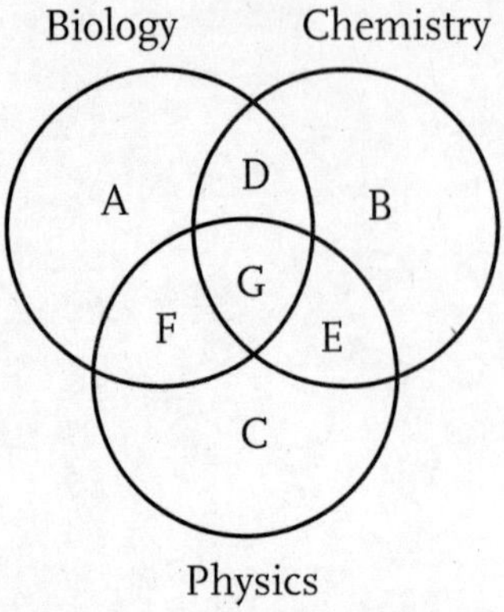

51. German Cars ( )

Choice C. There are two ways to solve this problem. One employs the "three-groups" formula, while the other involves using the Venn-diagram approach. The three-groups formula is preferable because it is clearly fastest, relying directly on the numbers found right in the problem. In short, the total number of individuals owning a single car are added together, the number of individuals owning exactly two of these cars is subtracted from this number, and, finally, the number of people owning all three of these cars are added back. The calculation follows:

Three-Groups Formula		Solve
Add:	BMW	45
Add:	Mercedes	38
Add:	Porche	27
Less:	BMW & Mercedes	<15>
Less:	BMW & Porche	<12>
Less:	Mercedes & Porche	<8>
Add:	BMW & Mercedes & Porche	5
Add:	None of BMW & Mercedes & Porche	0
	Total	80

Calculation:

$$B + M + P - BM - BP - MP + BMP + None = Total$$
$$45 + 38 + 27 - 15 - 12 - 8 + 5 + 0 = x$$
$$x = 80$$

The Venn-diagram approach is highly analytical and requires breaking the problem down into non-overlapping areas while finding individual values for all seven areas. That's right!—seven distinct areas are created when three groups overlap.

$$27 + 16 + 12 + 10 + 7 + 3 + 5 = 80$$

BMW only + Mercedes only + Porsche only – [(BMW & Mercedes) – (Mercedes & Porsche) – (BMW & Porsche)] + (BMW & Mercedes & Porsche).

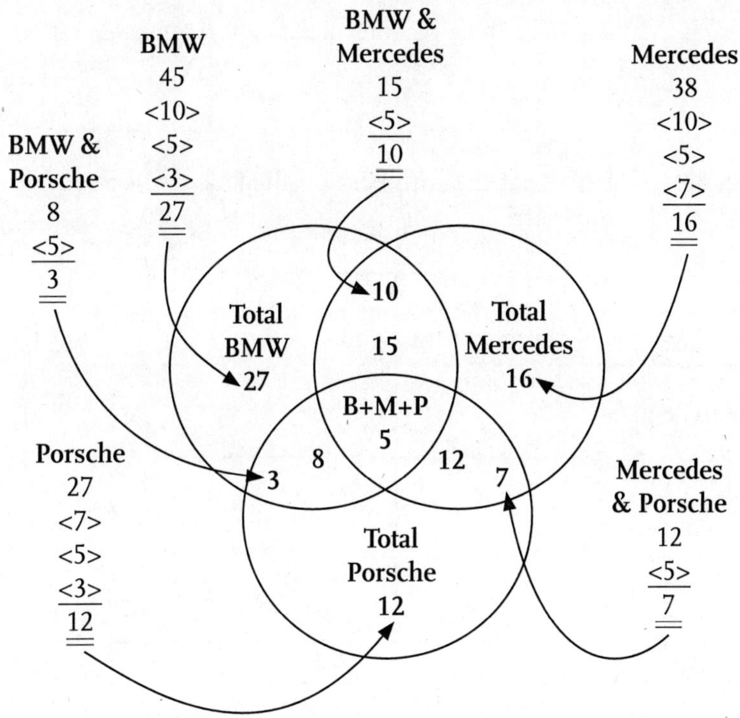

52. Job Search (🌶)

Choice C. Eleven of the candidates have at least seven years of work experience and hold degrees.

Step #1: Sketch a matrix and enter given information into the appropriate boxes. The question mark denotes the value we're trying to find.

	< 7 years of work experience	≥ 7 years of work experience	
With degrees		?	23
No degrees	3		
		20	35

Step #2: Let's total the numbers on the side and bottom of the matrix.

	< 7 years of work experience	≥ 7 years of work experience	
With degrees		?	23
No degrees	3		12
	15	20	35

Step #3: Since data must total down and across, we simply fill in remaining numbers within the middle four boxes.

	< 7 years of work experience	≥ 7 years of work experience	
With degrees	12	11	23
No degrees	3	9	12
	15	20	35

53. Nordic ()

Choice A. Five percent of the people in this Norwegian town have neither blond hair nor blue eyes. Note that since we are dealing with percentages, the total in the box in the bottom right-hand corner is 100%.

Step #1: Sketch a matrix and enter given information into the appropriate boxes.

	Blond Hair	Non Blond Hair	
Blue Eyes	45%		65%
Non Blue Eyes		?	
	75%		100%

Step #2: Total the numbers on the side and bottom of the matrix.

	Blond Hair	Non Blond Hair	
Blue Eyes	45%		65%
Non Blue Eyes		?	35%
	75%	25%	100%

Step #3: Since data must total down and across, we simply fill in remaining numbers within the middle four boxes.

	Blond Hair	Non Blond Hair	
Blue Eyes	45%	20%	65%
Non Blue Eyes	30%	5%	35%
	75%	25%	100%

54. Singles ()

Choice D. Two-thirds of the women are single (that is, $\frac{20}{30} = \frac{2}{3}$). For this problem, assume for simplicity's sake that there are 100 students in the course and fill in the given information, turning percentages into numbers. If 70% of the students are male then 30% must be female. If we assume there are 100 students then 70 are male and 30 are female. Note that if two-sevenths of the male students are married, then 20 male students are married; that is, two-sevenths of 70 male students equals 20 students.

First, plug the given data from the problem into the matrix:

	Male	Female	
Married	20		30
Single		?	
	70	?	

Second, fill in and complete the matrix:

	Male	Female	
Married	20	10	30
Single	50	20	70
	70	30	100

55. Batteries ( )

Choice A. To obtain the percentage of defective batteries sold by the factory, we fill in the information in the following matrix to obtain $\frac{3}{75}$ or $\frac{1}{25}$ or 4%. As in the previous problem, the technique of picking the number "100" greatly simplifies the task at hand.

First, plug the given data from the problem into the matrix:

	Defective	Not Defective	
Rejected		$\frac{1}{10}(80) = 8$	$\frac{1}{4}(100) = 25$
Not Rejected	?		?
	$\frac{1}{5}(100) = 20$	80	100

Second, fill in and complete the matrix:

	Defective	Not Defective	
Rejected	17	8	25
Not Rejected	3	72	75
	20	80	100

56. Experiment ( )

Choice A. Let's pick numbers. Say the total number of rats is 100, of which 60 are female and 40 are male. Let's say that 50 rats die, which means 15 were female and 35 were male. Calculation for dead rats: 30% × 50 = 15 female rats versus 70% × 50 = 35 male rats.

There are two ways to solve this problem: the "picking numbers" approach and the "algebraic" approach.

I. Picking Numbers Approach

Let's pick numbers. Say the total number of rats that are originally alive is 100, of which 60 are female and 40 are male. Let's say that 50 rats die, which means 15 were female and 35 were male. Calculation for dead rats is: 30% × 50 = 15 female rats versus 70% × 50 = 35 male rats.

	Male	Female	
Rats that Died	35	15	*50
Rats that Lived			
	40	60	100

	Male	Female	
Rats that Died	35	15	*50
Rats that Lived	5	45	**50
	40	60	100

Note that the figure "*50" is merely a guesstimate; the figure "**50" is a plug number. If there were 100 rats and 50 died, then 50 must have lived.

As it turns out, data about the number of rats that lived is not useful in the problem at hand; the focus is on the number of rats that died. We we are able to calculate the ratio of the death rate among male rats to the death rate among female rats as follows:

$$\frac{\frac{\text{male rat deaths}}{\text{male rats (total)}}}{\frac{\text{female rat deaths}}{\text{female rats (total)}}} = \frac{\frac{35}{40}}{\frac{15}{60}} = \frac{35}{40} \times \frac{60}{15} = 7:2$$

II. Algebraic Approach

$$\frac{\frac{0.7 \text{ died (male)}}{0.4 \text{ total (male)}}}{\frac{0.3 \text{ died (female)}}{0.6 \text{ total (female)}}} = \frac{0.7 \cancel{\text{died}}}{0.4 \cancel{\text{total}}} \times \frac{0.6 \cancel{\text{total}}}{0.3 \cancel{\text{died}}} = \frac{\cancel{0.42}^{7}}{\cancel{0.12}^{2}} = \frac{7}{2} = 7:2$$

The trap answer, per choice B, may be calculated in two ways. The first involves using numbers we previously obtained through the picking numbers approach. In this case: $\frac{35}{15} = \frac{7}{3} = 7:3$. However, this ratio is not correct because it involves dividing the estimated number of dead male rats by the number of dead female rats. We need to instead divide the death rate of male rats by the death rate of female rats.

Another way of obtaining trap answer choice B is as follows:

$$\frac{\frac{\text{male rat deaths}}{\text{male rats (total)}}}{\frac{\text{female rat deaths}}{\text{female rats (total)}}} = \frac{\frac{70\% \times 40\%}{40\%}}{\frac{30\% \times 60\%}{60\%}} = \frac{\frac{28\%}{40\%}}{\frac{18\%}{60\%}} = \frac{28\%}{40\%} \times \frac{60\%}{18\%} = 7:3$$

The pitfall here involves multiplying the respective death rates for male and female rats by the percentage of male and female rats. However, this is erroneous since we do not know how many rats died. In other words, we are dealing with two different groups of rats. The first group represents the total number of rats and the second group represents the rats that died. No direct link can be drawn between these two groups so we cannot simply multiply these percentages.

57. Petroleum ( )

Choice C. This is a wet mixture. We need to find the percentage of oil in a final solution.

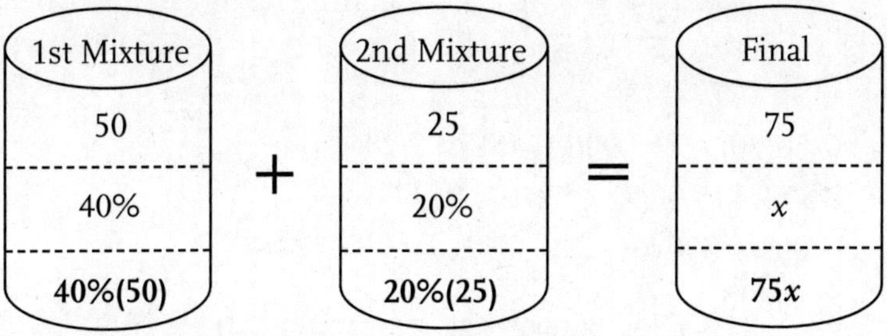

Next we create an equation from the totals (in bold) appearing at the bottom of the barrels:

$$40\%(50) + 20\%(25) = 75x$$
$$20 + 5 = 75x$$
$$25 = 75x$$
$$75x = 25$$
$$x = \frac{25}{75}$$
$$x = \frac{1}{3} \text{ or } 33\frac{1}{3}\%$$

58. Perfect Blend ( )

Choice B. This is a dry mixture. We need to calculate the amount of coffee in a second mixture that needs to be added to a first mixture to arrive at a final mixture.

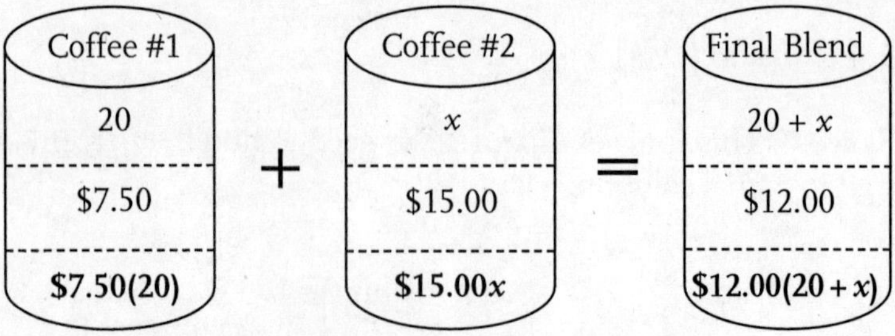

$\$7.50(20) + \$15.00(x) = \$12.00(20 + x)$

$\$150 + \$15.00(x) = \$240 + \$12.00x$

$\$3.00x = \$90$

$x = \dfrac{\$90}{\$3.00}$ (Note: dollar signs cancel)

$x = 30$ pounds

59. Nuts ( )

Choice A. This is a dry mixture. We need to calculate the amounts of two different nut mixtures to arrive at a final mixture.

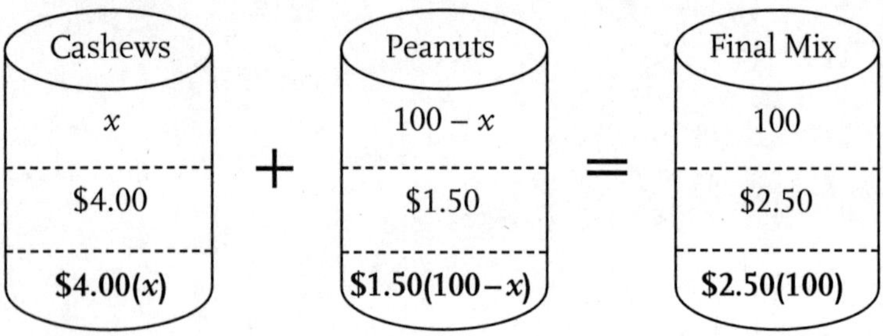

$\$4.00(x) + \$1.50(100 - x) = \$2.50(100)$

$\$4.00x + \$150 - \$1.50x = \$250$

$\$4.00x - \$1.50x = \$250 - \$150$

$\$2.50x = \$100$

$x = \dfrac{\$100}{\$2.50}$

$x = 40$

Therefore, the nut mixture consists of 40 pounds of cashews and 60 pounds of peanuts (that is, 100 – 40 = 60 pounds of peanuts).

60. Adding H_2O ( )

Choice B. This is a wet mixture. We need to calculate the amount of pure water that needs to be added to arrive at a final solution. Note that the percentage of water added is 0%. Pure water lacks any mixture. This causes the middle term in the equation to drop out.

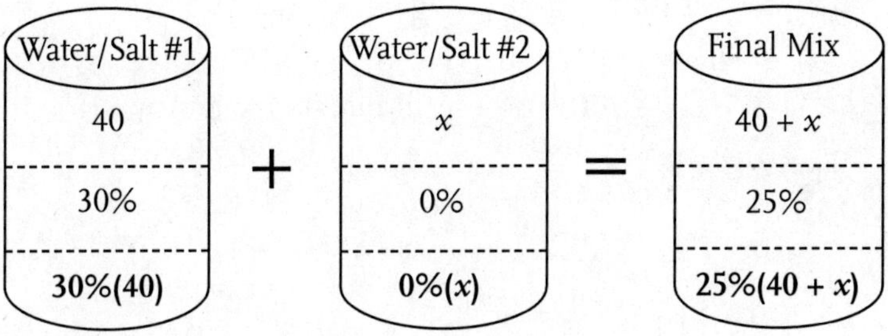

$$30\%(40) + 0\%(x) = 25\%(40 + x)$$
$$12 + 0 = 10 + 0.25x$$
$$2 = 0.25x$$
$$0.25x = 2$$
$$x = \frac{2}{0.25}$$
$x = 8$ gallons of water

61. Evaporation ( )

Choice E. This is a wet mixture. We need to calculate the amount of pure water that needs to be subtracted to arrive at a final solution. The percentage of the water subtracted is 0 percent. Pure water lacks any mixture. Again, this is the reason the middle term in the equation drops out.

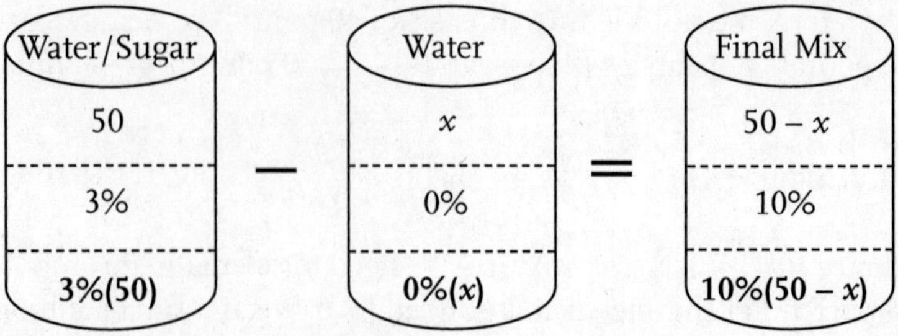

$3\%(50) - 0\%(x) = 10\%(50 - x)$

$1.5 - 0 = 5 - 0.10x$

$1.5 - 5 = -0.10x$

$-3.5 = -0.10x$

$(-1)(-3.5) = (-1)(-0.10x)$ [multiply both sides by –1]

$3.5 = 0.10x$

$0.10x = 3.5$

$x = \dfrac{3.5}{0.10}$

$x = 35$ liters of water

62. Gold ( )

Choice E. This is a dry mixture. We need to calculate the amount of pure gold that needs to be added to arrive at a final alloy.

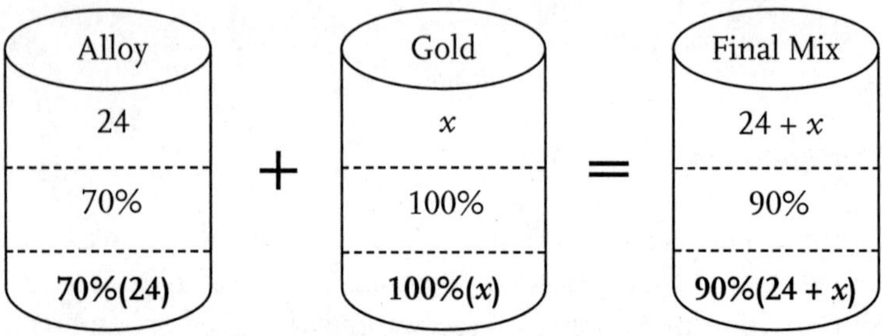

$70\%(24) + 100\%(x) = 90\%(24 + x)$

$16.8 + x = 21.6 + 0.90x$

$1.0x - 0.90x = 21.6 - 16.8$

$$0.10x = 4.8$$

$$x = \frac{4.8}{0.1}$$

$x = 48$ ounces of pure gold

NOTE When we add pure gold, we use 100%. Thus, the term at the bottom of the middle barrel becomes 100%(x). If we were to add a pure non-gold alloy, then the percentage would be 0% and the middle term would drop out in our calculation.

63. Bowling ( )

Choice B. Because the number of games bowled is equal, we can simply perform weighted average to arrive at the answer.

$$\text{Weighted Average} = (\text{Event}_1 \times \text{Weight}_1) + (\text{Event}_2 \times \text{Weight}_2)$$

$$\text{WA} = 110(60\%) + 130(40\%)$$

$$\text{WA} = 66 + 52$$

$$\text{WA} = 118$$

The alternative format below is more user-friendly when performing calculations.

$$\begin{array}{lr} \text{Event}_1 \times \text{Weight}_1 = xx & 110 \times 60\% = 66 \\ \text{Event}_2 \times \text{Weight}_2 = \underline{xx} & 130 \times 40\% = \underline{52} \\ \underline{\underline{xx}} & \underline{\underline{118}} \end{array}$$

64. The Right Mix ( )

Choice C.

$$\text{Weighted Average} = (\text{Event}_1 \times \text{Weight}_1) + (\text{Event}_2 \times \text{Weight}_2)$$

$$WA = \$900(80\%) + \$15.00(20\%)$$
$$WA = \$7.20 + \$3.00$$
$$WA = \$10.20$$

This is de facto a weighted average problem. As such, we simply multiply dollars (events) by their respective percentages (weights).

65. Baseball ( )

Answer A.

$$\text{Weighted Average} = (\text{Event}_1 \times \text{Weight}_1) + (\text{Event}_2 \times \text{Weight}_2)$$

$$WA = (3\% \times 0.5) + (5\% \times 0.5)$$
$$WA = 1.5\% + 2.5\%$$
$$WA = 4\%$$

If the price of one baseball bat increases by 3%, then the price of three baseball bats also increases by 3% (not 9% or 12%). The percentage change in aggregate price (cost) is the same regardless of the number of baseball bats in question. Ditto for a catcher's mitt.

Since a baseball bat and a catcher's mitt currently cost the same, the baseball bat and catcher's mitt each cost one-half of the total price. So one half of the total price has increased by 3%, while the other half has increased by 5%. Therefore, the total price has increased by 4%.

66. Venture Capital ( )

Choice B. Expected return or expected earnings is calculated using the weighted average formula for three events.

The weighted average formula can be summarized as follows:

$$WA = (E_1 \times W_1) + (E_2 \times W_2) + (E_3 \times W_3)$$

$$WA = \$5{,}000{,}000(20\%) + \$1{,}500{,}000(50\%) + \$500{,}000(30\%)$$

$$WA = \$1{,}000{,}000 + \$750{,}000 + \$150{,}000$$

$$WA = \$1{,}900{,}000$$

The following format is "computation-friendly":

$$\begin{aligned} \$5{,}000{,}000 \times 20\% &= \$1{,}000{,}000 \\ \$1{,}500{,}000 \times 50\% &= \$750{,}000 \\ 130 \times 40\% &= \underline{\$150{,}000} \\ & \underline{\underline{\$1{,}500{,}000}} \end{aligned}$$

67. Portfolio ( )

Choice A. Weighted average, again, is the basis for calculating the expected return.

$$\text{Expected Return} = WA_1 \times WA_2 + WA_3$$

First Investment:

$$WA_1 = (E_1 \times W_1) + (E_2 \times W_2) + (E_3 \times W_3)$$

$$WA_1 = \$90{,}000\left(\tfrac{1}{6}\right) + \$50{,}000\left(\tfrac{1}{2}\right) + -\$60{,}000\left(\tfrac{1}{3}\right)$$

$$WA_1 = 15{,}000 + 25{,}000 + -20{,}000$$

$$WA_1 = 20{,}000$$

Second Investment:

$$WA_2 = (E_1 \times W_1) + (E_2 \times W_2)$$

$$WA_2 = \$100,000\left(\frac{1}{2}\right) + -\$50,000\left(\frac{1}{2}\right)$$

$$WA_2 = \$50,000 + -25,000$$
$$WA_2 = 25,000$$

Third Investment:

$$WA_3 = (E_1 \times W_1) + (E_2 \times W_2) + (E_3 \times W_3) + (E_4 \times W_4)$$

$$WA_3 = \$100,000\left(\frac{1}{4}\right) + \$60,000\left(\frac{1}{4}\right) + (-\$40,000)\left(\frac{1}{4}\right) + (-\$80,000)\left(\frac{1}{4}\right)$$

$$WA_3 = \$25,000 + \$15,000 + -\$10,000 + -\$20,000$$
$$WA_3 = 10,000$$

So, the expected return on all three investments is:

Expected Return = \$20,000 + \$25,000 + \$10,000
Expected Return = \$55,000

Chapter 3 – Favorite Numeracy Dishes

68. Gross Margin ( )

Choice B.

$$\text{Margin \%} = \frac{\text{Sales (\$)} - \text{Cost (\$)}}{\text{Sales (\$)}} \times 100\%$$

$$\text{Margin \%} = \frac{\$10 - \$7.50}{\$10} \times 100\%$$

$$\text{Margin \%} = \frac{\$2.50}{\$10} \times 100\%$$

$$\text{Margin \%} = 0.25 \times 100\% = 25\%$$

69. Markup ( )

Choice C.

$$\text{Markup}\,\% = \frac{\text{Sales (\$)} - \text{Cost (\$)}}{\text{Cost (\$)}} \times 100\%$$

$$\text{Markup}\,\% = \frac{\$10 - \$7.50}{\$7.50} \times 100\%$$

$$\text{Markup}\,\% = \frac{\$2.50}{\$7.50} \times 100\%$$

$$\text{Markup}\,\% = 0.33 \times 100\% = 33\tfrac{1}{3}\%$$

70. Alpha Inc. ( )

Choice C.

$$\text{Margin}\,\% = \frac{\text{Markup}\,\%}{100\% + \text{Markup}\,\%}$$

$$\text{Margin}\,\% = \frac{50\%}{100\% + 50\%}$$

$$\text{Margin}\,\% = \frac{50\%}{150\%}$$

$$\text{Margin}\,\% = 33\tfrac{1}{3}\%$$

NOTE ☞ The calculation below is technically more accurate. The percentage signs (that is, %) cancel out and 0.33 must be multiplied by 100% in order to turn this decimal into a percentage and to reinstate the percentage sign in the final answer.

$$\text{Margin}\,\% = \frac{50\%}{150\%} = 0.33 \times 100\% = 33\tfrac{1}{3}\%$$

71. Beta Corp. ( )

Choice B.

$$\text{Markup}\% = \frac{\text{GM}\%}{100\% - \text{GM}\%}$$

$$\text{Markup}\% = \frac{20\%}{100\% - 20\%}$$

$$\text{Markup}\% = \frac{20\%}{80\%}$$

$$\text{Markup}\% = 25\%$$

Again, the following calculation is more accurate. The percentage signs (that is, %) cancel out and 0.25 must be multiplied by 100% in order to turn this decimal into a percentage and to reinstate the percentage sign in the final answer.

$$\text{Markup}\% = \frac{20\%}{80\%} = 0.25 \times 100\% = 25\%$$

72. Century Bus Line ( )

Choice D. Profit is the interplay of three variables—price, cost, and volume. The key is to review the profit formula to determine which variables are increasing or decreasing in value:

$$(\text{Price per unit} - \text{Cost per unit}) \times \text{No. of units} = \text{Profit}$$

Maximum profit will be obtained when price per unit goes up, costs per unit go down, and sales volume goes up. This may be summarized as follows:

$$(\uparrow\text{Price per unit} - \downarrow\text{Cost per unit}) \times \uparrow\text{No. of units} = \uparrow\uparrow\text{Profit}$$

In this problem, we know that volume (that is, the number of paying passengers) is forecasted to increase. Therefore, we must look at the unknown variables, namely price per unit (that is, ticket sales price per passenger) and total cost per unit (that is, cost per passenger) as follows:

$$(?\,\text{Price per unit} - ?\,\text{Cost per unit}) \times \uparrow\text{No. of units} = \downarrow\text{Profit}$$

In summary, either the price per unit of ticket sales will decrease or total costs per unit will increase. No answer choice gives us the option of an increase in costs. Therefore, it must be that price per unit (ticket sales price per passenger) will go down significantly to more than offset the increase in sales volume. A significant reduction in the price of fares would have the effect of reducing overall profits, and 25% would be considered a significant reduction. Answer choices A, B, and C are essentially irrelevant to this scenario as well, and answer choice E would actually serve to increase forecasted profits.

73. High-Speed Boat ( )

Choice B. The cost is $g + h$ cents per mile. We must multiply this by 100 miles and divide by 100 cents to get dollars. Doing so will result in the original $g + h$.

$$\text{Total Cost} = (\text{Cost}_1 + \text{Cost}_2) \times \text{Number of units}$$

$$\text{Total Cost} = (g + h)\frac{\text{cents}}{\cancel{\text{miles}}} \times 100\ \cancel{\text{miles}}$$

$$\text{Total Cost} = 100(g + h)\ \text{cents}$$

$$\text{Total Cost} = \cancel{100}(g + h)\ \cancel{\text{cents}} \times \frac{1\ \text{dollar}}{\cancel{100}\ \cancel{\text{cents}}}$$

$$\text{Total Cost} = (g + h)\ \text{dollars}$$

74. Cases (🌶🌶)

Choice B. The profit formula may be manipulated or "rearranged" in order to solve for revenue. Note that the price charged may be viewed as total revenue.

$$\text{Revenue} - \text{Cost} = \text{Profit}$$
$$\text{Revenue} = \text{Cost} + \text{Profit}$$
$$\text{Revenue} = \frac{1{,}000(5)}{100} + \frac{9{,}000(y)}{100} + x$$
$$\text{Revenue} = 50 + 90y + x$$

NOTE It is easy, when doing these calculations quickly, to forget not only to add x at the end (which represents the expected profit) but also to divide cost components by 100 in order to convert cents to dollars.

75. Delicatessen (🌶🌶)

Choice B. A variation of the profit formula is:

$$\text{Profit} = (\text{Price per unit} \times \text{No. of units}) - (\text{Cost per unit} \times \text{No. of units})$$

$$\text{Profit} = s\,\frac{\text{dollars}}{\text{pound}} \times \left(p \text{ pds} - d \text{ pds}\right) - \left(c\,\frac{\text{dollars}}{\text{pound}} \times p \text{ pds}\right)$$

$$\text{Profit} = s(p - d) - cp$$

NOTE Gross profit is sales revenue minus product cost. Sales revenue is price per unit multiplied by the number of units: $s(p - d)$. Product cost is cp. Therefore, gross profit is $s(p - d) - cp$.

76. Paper Clips ( )

Choice A. In this problem, cost per unit is c/p. The volume or unit figure is q. Total cost is the product of these two variables.

$$\text{Total Cost} = \text{Cost per unit} \times \text{Number of units}$$

$$\text{Total Cost} = \frac{c \text{ cents}}{p\ \cancel{\text{clips}}} \times q\ \cancel{\text{clips}}$$

$$\text{Total Cost} = \frac{cq}{p} \text{ cents}$$

77. Garments ( )

Choice B. This problem requires us to calculate number of units or volume. Start with the basic cost formula and solve for number of units as follows:

$$\text{Total Cost} = \text{Cost per unit} \times \text{Number of units}$$

$$\text{Therefore, Number units} = \frac{\text{Total Cost}}{\text{Cost}_{\text{ per unit}}}$$

$$\text{Number of units} = \frac{t \text{ dollars}}{\dfrac{d \text{ dollars}}{s \text{ shirts}}}$$

$$\text{Number of units} = t\ \cancel{\text{dollars}} \times \frac{s \text{ shirts}}{d\ \cancel{\text{dollars}}}$$

$$\text{Number of units} = \frac{ts}{d} \text{ shirts}$$

In summary, since the price per unit equals d/s, we divide t by d/s to arrive at ts/d.

78. Copy ( )

Choice C.

$$\text{Total Cost} = \text{Cost per unit} \times \text{Number of units}$$

$$\text{Total Cost} = 2\,\frac{\text{cents}}{\text{page}} \times (y \text{ pages} \times y \text{ pages})$$

$$\text{Total Cost} = 2\,\frac{\text{cents}}{\cancel{\text{page}}} \times y^2\ \cancel{\text{pages}}$$

$$\text{Total Cost} = 2y^2 \text{ cents}$$

Another approach is to first calculate the cost of each manuscript and multiply this expression by the number of manuscripts. Thus, the cost to copy one manuscript ($2y$), is multiplied by the number of manuscripts (y) in order to yield the cost of all manuscripts—$2y^2$. That is, 2 cents per page times y number of pages per manuscript times y manuscripts: $2y(y) = 2y^2$.

79. Pete's Pet Shop ( )

Choice B. This problem requires that we calculate a usage per unit figure.

$$\text{Cups per bird per day} = \frac{35 \text{ cups}}{15 \text{ birds}} \times \frac{1}{7 \text{ days}}$$

$$\text{Cups per bird per day} = \frac{1}{3} \text{ of a cup per bird per day}$$

Therefore, the number of cups of bird seed to feed 9 birds for 12 days is:

$$\frac{1}{3} \text{ of a cup per bird per day} \times 9 \text{ birds} \times 12 \text{ days} = 36 \text{ cups}$$

80. Book Publisher ( )

Choice E. This is a classic break-even problem. Note that this problem is not asking for how much profit but rather for the break-even point (that is, the point of zero profit, achieved when variable revenue is exactly equal to fixed costs or expenses). Moreover, fixed costs and variable costs differ in that variable costs vary directly with sales whereas fixed costs are the same regardless of whether you sell a single item or one million items.

Algebraic Method:

Let x be the number of units. The algebraic method is based on a variation of the break-even point formula. Here the break-even point occurs exactly where total costs equal total revenue. Total costs equal fixed costs plus variable costs; total revenue equals sales price multiplied by the number of units sold. By setting total revenue equal to total costs, we arrive at the break-even point.

Total Revenue = Total Costs

Total Revenue = Fixed Costs + Variable Costs

Total Revenue = FC + VC per unit × No. of units)

Price per unit × No. of units = FC + (VC per unit × No. of units)

Calculation:

$$\$5.95x = (\$150{,}000 + \$20{,}00 + \$230{,}000) + (0.33 + 0.12 + 0.50)x$$

$$\$5.95x = \$150{,}000 + \$20{,}00 + \$230{,}000 + 0.33x + 0.12x + 0.50x$$

$$\$5.95x = \$400{,}000 + \$0.95x$$

$$\$5.95x - \$0.95x = \$400{,}000$$

$$\$5.00x = \$400{,}000$$

$$x = \frac{\$400{,}000}{\$5.00}$$

$x = 80{,}000$ copies (or units)

Accountant's Method (per Tip #21, *Chapter 3*):

$$BE_{units} = \frac{\text{Fixed Costs}}{\text{Selling Price}_{\text{per unit}} - \text{Variable Cost}_{\text{per unit}}}$$

Calculation:

$$\text{Break Even}_{units} = \frac{\$400{,}000}{\$5.95 - \$0.95}$$

$$\text{Break Even}_{units} = \frac{\$400{,}000}{\$5.00}$$

$$\text{Break Even}_{units} = 80{,}000 \text{ copies (units)}$$

NOTE When dividing $400,000 by $5 per copy, the dollar signs cancel, leaving copies (or units).

81. Sabrina to Change Jobs ( )

Choice C. The difference between Sabrina's current base salary, $85,000, and $45,000 is $40,000. Divide $40,000 by 15%($1,500) to get 177.77 unit sales. In the equation below, x stands for the number of sales.

$$\text{Revenue}_{\text{Option 1}} = \text{Revenue}_{\text{Option 2}}$$

$$\$85{,}000 = \$45{,}000 + 0.15(\$1{,}500)(x)$$

$$\$85{,}000 - \$45{,}000 = 0.15(\$1{,}500)(x)$$

$$\$40{,}000 = 0.15(\$1{,}500)(x)$$

$$\$40{,}000 = \$225(x)$$

$$\$225(x) = \$40{,}000$$

$$x = \frac{\$40{,}000}{\$225}$$

$$x = 177.77$$

Therefore, $x = 178$ unit sales

NOTE ✍ When dividing $40,000 by $225 (commission) per sale, the dollar signs cancel, leaving number of sales (or units).

Don't be tricked by tempting wrong answer choice B. A total of 177 sales isn't enough to break even. This number must be rounded up to 178 in order to avoid losing money. The actual number of sales is discrete, and can only be represented by whole numbers, not decimals.

With reference to Tip #21 *(Chapter 3)*, the $40,000 in this problem may be viewed as the fixed costs while the $225(0.15 × $1,500) figure may be viewed as the variable revenue per unit. Thus, $40,000 divided by $225 is equal to 177.77 in unit sales.

82. Grapes (🌶🌶)

Choice B. The assumption is that both crops, grapes and watermelons, were grown on the same number of acres. Of course, per acre costs in each U.S. state are determined as follows:

$$\frac{\text{Costs}}{\text{Acres}} = \text{Cost per acre}$$

In choice B, we cannot make an assessment on the cost per acre basis without knowing the total costs and the total acres. Despite the greater total cost to grow watermelons in Oklahoma (compared with grapes in California), if many more acres of watermelons were grown in Oklahoma (as compared with grapes in California), then the cost per acre to grow watermelons in Oklahoma might very well be lower.

Answer choices A and C fall outside the scope of the argument. In choice A, we are talking about costs to grow, not profits on the sale. Choice C switches terms around. In the original, we are only talking about watermelons in Oklahoma and grapes in California. We must stick with the information as presented. In choice D, how the crops are being used is irrelevant. In choice E, the source of money used to grow crops is also irrelevant. It does not matter who pays for the cost of growing the crops—farmers, the government (state subsidies), or Aunt Jessie—costs are costs.

83. Act-Fast ( )

Choice D. This problem is built on the assumption that the number of tablets per bottle is equal. Therefore, if a bottle of regular aspirin contains more than twice as many tablets as does a bottle of Act-Fast, then logic dictates that a bottle of regular aspirin provides a greater amount of pain reliever. And since the cost per bottle is the same, it stands to reason that the cost efficiency of regular aspirin is greater than that of Act-Fast.

What we really want to determine is how much pain reliever we're getting and how much it is costing us. In choice A, there is no mention of the number of tablets per bottle, so we cannot determine how much pain reliever in aggregate we're getting for our money. Furthermore, implied logic dictates that if one Act-Fast tablet contains twice the pain reliever found in a tablet of regular aspirin then an Act-Fast tablet should be twice the size of a regular aspirin tablet (or an aspirin tablet should be one-half the size of an Act-Fast tablet). Whereas choice B is beyond the scope of the argument, choice C weakens the argument but not nearly to the extent choice D does. There is no need for the pain reliever to be of a different type or strength, as suggested by choice E.

84. Prototype ( )

Choice B. This problem highlights the concepts of "efficiency" and "cost efficiency."

First, we set up the problem conceptually:

$$\left(1.0\,\frac{\text{dollar}}{\cancel{\text{gallon}}} \times 1.0\,\cancel{\text{gallons}}\right) - \$x = \left(1.2\,\frac{\text{dollar}}{\cancel{\text{gallon}}} \times \frac{5}{9}\,\cancel{\text{gallons}}\right)$$

Remember that $x is equal to the dollar savings. Question: Where does the fraction $\frac{5}{9}$ come from? An 80 percent increase in efficiency can be expressed as $\frac{180\%}{100\%}$ or $\frac{9}{5}$. The reciprocal of $\frac{9}{5}$ is $\frac{5}{9}$. This means that the P-Car needs only five-ninths as much fuel to drive the same distance as does the T-Car.

Second, we solve for *x*, which represents the cost savings when using the P-Car.

Let's convert \$1.2 to $\frac{6}{5}$ for simplicity.

$$\$(1.0)(1.0) - \$x = \$\left(\frac{6}{5}\right)\left(\frac{5}{9}\right)$$

$$\$1.0 - \$x = \$\left(\frac{6}{\cancel{5}}\right)\left(\frac{\cancel{5}}{9}\right)$$

$$\$1 - \$x = \$\frac{2}{3}$$

$$-\$x = -\$1 + \$\frac{2}{3}$$

$$-\$x = -\$\frac{1}{3}$$

$$(-1)(-\$x) = (-1)(-\$\tfrac{1}{3}) \quad \text{[multiply both sides by } -1]$$

$$\$x = \$\frac{1}{3} \quad \text{[dollar signs cancel]}$$

$$x = \frac{1}{3} \text{ or } 33\tfrac{1}{3}\%$$

Since we save one-third of a dollar for every dollar spent, our percentage savings is $33\frac{1}{3}$ percent. Therefore, although the cost of gas per gallon for the P-Car is more expensive, it results in an overall cost efficiency.

For the record, whereas 80 percent represents how much more efficient per gallon the P-Car is compared to the T-Car, the correct answer, $33\frac{1}{3}$ percent represents how much more cost efficient per gallon the P-Car is compared to the T-Car.

85. Wanda ( )

Choice C. If we assume that the problems on the test are distributed proportionally or linearly across the four sections of the test, then yes, Wanda has finished more than 75 percent of the test. She has finished 76 problems out of 100 and is now working on her 77th problem. If each of the four test sections do not contain the same number of problems, then we cannot tell whether she has finished more or less than 75 percent of the exam.

86. Noble Book Club ( )

Choice B. The argument claims that the free CD-ROM gift program has succeeded in increasing Noble Book Club sales. However, this claim would be greatly weakened if the company sold the exact same number of books, but just had more concentrated sales patterns at specific points in time. In other words, instead of 100 people buying one book each, perhaps 25 people bought four books each. The free CD-ROM gift program may have succeeded in selling more books to selected customers, but overall, there might be little change in terms of the total number of books sold.

The sales representative's assumption is that the free CD-ROM gift program is the driving force behind increased sales. Choice A would fall outside the scope of the argument. We are concerned in this argument with the linkage between the CD-ROM gift and sales of four or more books, not with sales of less than four books. Choice C is irrelevant since we are only concerned about the increase in sales, not costs, and not profits. Choice D is irrelevant because what other book clubs did or did not do is of little concern. Choice E only says membership is up in general, but that is not the same thing as saying Noble Book Club sales increased.

87. Max Motorcycles (🌶🌶)

Choice D. If the number of motorcycles built by Max Motorcycles has not increased sharply in the past 10 years, then this will strengthen the argument. But if the number of motorcycles built by Max has increased sharply in the past 10 years, then this will weaken the argument. For it could be that a large number of Max Motorcycles that are on the road today were just recently manufactured and there hasn't been sufficient time to test their durability. Therefore, by knowing that this is not the case, the argument is strengthened. This concept is referred to as back-loading as opposed to front-loading. The argument is strengthened knowing that back-loading (a recent increase in production of motorcycles) is not occurring. Stated another way, the argument is weakened knowing that back-loading is occurring, that is, a large proportion of new motorcycles has been recently released in the marketplace.

Chapter 4 – Special Math Garnishments

88. Quiz on Basic Graphs (🌶)

Rising Graphs (select from graphs A, B, C or D):

You are paid incrementally more dollars for each hour you work.	B
You get a fixed base salary but may also get additional pay based on hours worked which are not covered by your base salary.	D
You get paid incrementally fewer dollars for each hour you work.	C
For every hour you work, you get paid the same number of dollars per hour.	A

Flat Graphs (select from graphs E, F, G or H):

You got a promotion at work that increased the fixed salary you now receive.	F
Because of layoffs at work, you decided to take a reduction in your fixed monthly salary in order to keep your current job.	G
You get paid a fixed (flat) wage regardless of the hours you work.	E
You get paid by the hour, but cannot earn more than a certain dollar amount during the course of your working week.	H

Falling Graphs (select from graphs I, J, K or L):

Your yearly pension decreases in terms of real dollars during each year of your retirement.	I
Your wages decrease gradually with age until the time you stop working altogether.	K
Your wages are fixed but stop the moment you retire.	L
Your wages decrease dramatically with age but you can still earn a little in your retirement.	J

89. Grand Hotel Q1 ( )

Answer: $4,000,000. We know that card and table games revenue totals $1,000,000 and represents 25% of the pie. For example, total revenue from all card and table games—blackjack ($200,000), roulette ($500,000), craps ($100,000), and baccarat ($200,000)—amounted to $1,000,000 (see bar chart).Therefore, this relationship provides the key to solving for the total amount of money brought in by Grand Hotel & Casino.

$$\frac{25\%}{\$1{,}000{,}000} = \frac{100\%}{x}$$

$$25\%(x) = 100\%(\$1{,}000{,}000)$$
$$25\%(x) = \$1{,}000{,}000$$
$$x = \frac{\$1{,}000{,}000}{25\%}$$
$$x = \$4{,}000{,}000$$

90. Grand Hotel Q2 (🌶🌶)

Answer: 5%. Blackjack is a type of card and table game. Blackjack accounted for \$200,000 worth of total card & table game revenue. Again, total revenue from all card & table games, per bar chart, totaled \$1,000,000 (blackjack, \$200,000; roulette, \$500,000; craps, \$100,000; and baccarat, \$200,000). Thus, blackjack accounted for 20% of card & table game revenue (\$200,000/\$1,000,000 = 20%). Since the percentage of total hotel and casino revenue attributable to card and table games was 25% (see pie chart), the final calculation is the product of these two numbers (percentages).

$$20\% \times 25\% = 5\%$$

NOTE ✍ With respect to bar charts and pie charts used in combination, concentrate on how one graph fits into the other. Because pie charts are best at breaking down information and bar charts are best at ranking information, the total dollar amount represented by the bar chart will most likely relate to a given percentage of the pie chart.

91. Grand Hotel Q3 (🌶)

Answer: miscellaneous revenues. The amount of money earned by craps was \$100,000. This figure can be easily read from the bar chart. The amount of money earned from miscellaneous revenues was 3% of \$4,000,000 or \$120,000. Again, the \$4,000,000 figure, as derived in Q1, represents the total dollar revenue earned by Grand Hotel & Casino.

92. Workforce Q1 (🌶)

Answer: 1.5 million. First, the number of people working in the blue-collar sector in 1980 was 4.5 million (45% × 10,000,000 = 4.5 million). Second, the number of people working in the blue-collar sector in 2010 was 3 million (20% × 15,000,000 = 3 million). Therefore, the difference between 4.5 million and 3 million accounts for the 1.5 million decrease in the number of workers in this sector.

NOTE ☙ When viewing two pie charts and comparing the percentage growth of items over two different time frames, beware of directly comparing the percentage figures of one pie chart with the other. To compare two figures, first translate these figures to actual dollars, based on the dollar size of each pie chart, and then make dollar or percentage comparisons.

93. Workforce Q2 (🌶🌶)

Answer: The Blue-Collar Sector. The number of people working in the service sector in 2010 was 4.5 million (30% of 15,000,000). The number of people working in the blue-collar sector in 1980 was also 4.5 million (45% of 10,000,000). Although the absolute percentage of service sector workers in 2010 is smaller than the absolute percentage of blue-collar workers in 1980, the size of the workforce has grown by 5 million. The pie has gotten bigger. In this case, a smaller percentage of a bigger pie is equal to a larger percentage of a smaller pie.

94. Workforce Q3 ( )

Answer: Service, Professional, and Clerical Sectors. All three sectors experienced a higher percentage increase compared with the percentage decrease attributed to the blue-collar sector!

The tricky point here is that we cannot compare percentages directly. In other words, we cannot simply subtract the percentages for any particular sector to arrive at a percentage increase or decrease. For example, the absolute percentage decrease in the number of workers in the blue-collar sector (that is, 45% – 20% = 25%) is obviously larger than any absolute percentage increase for any other sector. However, this type of calculation is not possible because we're asked for the percentage increase or decrease in the number of workers in each sector. These are relative numbers and we must take into account the fact that the size of the pies we're comparing are not the same. The workforce was larger in 2010.

Service Sector (percentage increase):

$$\frac{\text{New} - \text{Old}}{\text{Old}} \qquad \frac{4.5\text{M} - 1.5\text{M}}{1.5\text{M}}$$

$$\frac{3\text{M}}{1.5\text{M}} = 2 \times 100\% = 200\%$$

Professional Sector (percentage increase):

$$\frac{\text{New} - \text{Old}}{\text{Old}} \qquad \frac{3.75\text{M} - 1.5\text{M}}{1.5\text{M}}$$

$$\frac{2.25\text{M}}{1.5\text{M}} = 1.5 \times 100\% = 150\%$$

Clerical Sector (percentage increase):

$$\frac{\text{New} - \text{Old}}{\text{Old}} \qquad \frac{3.75\text{M} - 2.5\text{M}}{2.5\text{M}}$$

$$\frac{1.25\text{M}}{2.5\text{M}} = 0.5 \times 100\% = 50\%$$

Blue-Collar Sector (percentage decrease):

$$\frac{\text{Old} - \text{New}}{\text{Old}} \qquad \frac{4.5\text{M} - 3\text{M}}{4.5\text{M}}$$

$$\frac{1.5\text{M}}{4.5\text{M}} = \frac{1}{3} \times 100\% = 33\frac{1}{3}\%$$

95. Magna Fund Q1 ( )

Choice C. Refer to the chart titled *Value of Portfolio.* The y-axis is labeled in increments of \$50 million. We can estimate the value of the portfolio during Q1 as being \$75 million. The value of the portfolio during Q3 sits at \$125 million. The difference between these two segments is \$50 million. Therefore, \$50 million divided by \$75 million is equal to a $66\frac{2}{3}\%$ increase.

96. Magna Fund Q2 ( )

Choice C. Statement I is incorrect. The percentage of cash increases from the third to fourth quarter (Q3 to Q4). Although we see a black downward-sloping line, don't be deceived. This means that the percentage of cash is getting larger. In Q4, cash amounts to fully 50% of the value of the portfolio. When reading a cumulative line graph, concentrate on the space or area covered by each item in the cumulative line graph rather than on whether the lines enclosing these given areas are pointing upward or downward.

Statement II is also incorrect. Bonds accounted for one-third or more of the portfolio in exactly two of four quarters, namely the III and IV quarters. Note that one-third is $33\frac{1}{3}$%, not 30%; bonds only accounted for exactly 30% of the portfolio in Q1, not percent.

Statement III is correct. The average amount of dollar investment per individual investor was greatest during the fourth quarter. Don't be deceived by the fact that Q3 has the greatest portfolio value and the greatest number of investors. Average means total dollars divided by total investors. So lots of money divided by fewer investors means rising investment per individual investor. Throughout the first, second, and third quarters, the investment per individual investor remains relatively stable. In the fourth quarter the value of the portfolio falls and so does the number of investors. However, the number of investors falls proportionately greater, meaning that more money is divided among fewer investors. Actual or approximate numbers are as follows:

Quarter 1:

$$\frac{\$75{,}000{,}000}{85{,}000} = \$882 \text{ per investor}$$

Quarter 2:

$$\frac{\$100{,}000{,}000}{100{,}000} = \$1{,}000 \text{ per investor}$$

Quarter 3:

$$\frac{\$125{,}000{,}000}{110{,}000} = \$1{,}136 \text{ per investor}$$

Quarter 4:

$$\frac{\$75{,}000{,}000}{40{,}000} = \$1{,}875 \text{ per investor}$$

Statement IV is also correct. In the second quarter, stocks account for 60% of the total portfolio (80% less 20%). This means that preferred shares are 30% of 60%, or preferred shares accounted for 18% of the total portfolio. Looking at the *Value of Portfolio* chart, we can see that total portfolio value during Q2 is $100,000,000.

The final calculation follows: 18% × $100,000,000 = $18,000,000. Alternatively, you may prefer working from largest to smallest: ($100,000,000 × 60%) × 30% = $18,000,000.

97. Quiz on Correlation Analysis ( )

	Graph
Negative-high correlation	A
Positive-low correlation	B
Negative-perfect correlation	C
Zero correlation	D
Positive-high correlation	E
Negative-low correlation	F
Positive-perfect correlation	G
Non-linear correlation	H

98. Trends Q1 ( )

Answers: Graphs A and B. The graphs of which individual products show an identical mean, median, and mode?

Both graphs of products A and B have identical means, medians, and modes, all of which are equal to 50 years of age. Note the dotted vertical line that intersects the top of each graph. A graph that is completely symmetrical will have an identical mean, median, and mode.

99. Trends Q2 ( )

Answer: Graph C. The graph of which of the four products shows the smallest mode?

Product C has the smallest mode while product D has the largest mode. The mode is the place where each graph peaks because this is the spot where a value appears most frequently. For the record, product C has a mode of 35, product A has a mode of 50, product B has a mode of 50, and product D has a mode of 65.

100. Trends Q3 ( )

Answer: Graph C. The graph of which product shows a greater mean compared with its mode?

Visually, the graph of product C is skewed toward the larger values, and the mean is more affected by these larger values than is the median or mode. In statistical parlance, graphs skewed to the right (with a shape similar to the graph below) will always have a mean that is greater than their median and a median which is, in turn, greater than their mode (that is, mean > median > mode).

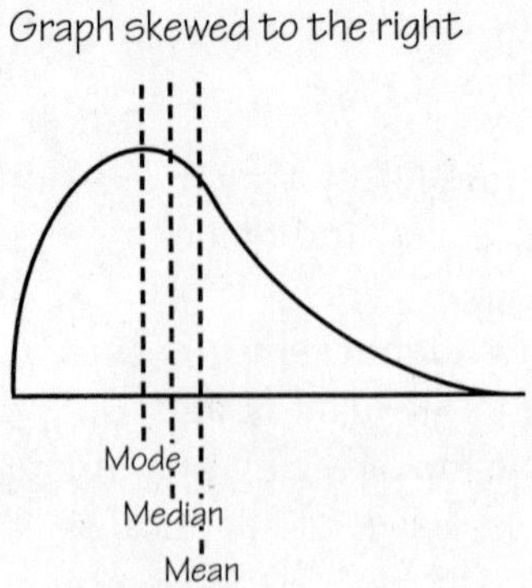

A classic real-life situation of a graph skewed to the right (toward larger values) is that of population and income earnings. Most people earn about the same, but a very small percentage of the population earns a tremendous amount, and this discrepancy pulls the mean toward these larger values. The median is also pulled toward these larger values, but less dramatically so.

NOTE ✍ Graphs that are skewed to the right have their tails on the right-hand side. They are so called because they taper off to the right. Visually, the highest point on the graph is situated to the left, and this may make the label "skewed to the right" seem counterintuitive. The exact opposite holds true for graphs that are skewed to the left, as seen in the next pictorial.

101. Trends Q4 (🌶🌶)

Answer: Graph D. The graph of which product shows a greater mode compared with its respective mean?

Visually, the graph of product D is skewed left (toward smaller values), and the mean is more affected by the inclusion of these smaller values than is the median or mode. In statistical parlance, graphs skewed to the left (with a shape similar to the graph below) will always have a mode that is greater than their median, which is, in turn, greater than their mean (that is, mean < median < mode).

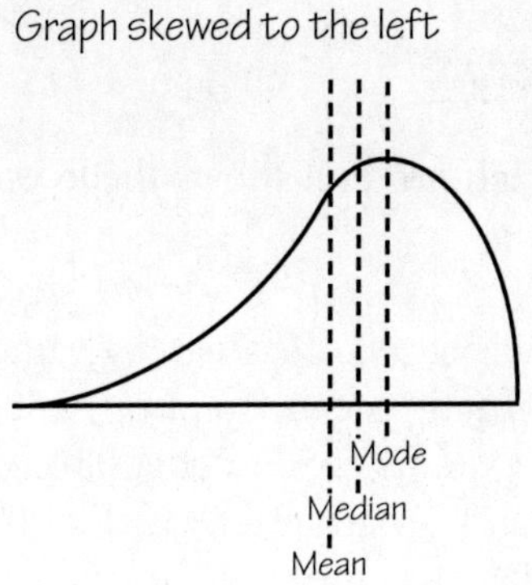

Graphs skewed to the left are less common in real life than are graphs which are skewed to the right. An example of a graph that is skewed to the left would be the date printed on coins. If we were to go looking for coins, most of those we'd find in circulation would be newly minted. If we looked hard enough, eventually we'd find some very old coins. The number of old coins would create a graph skewed to the left (that is, number of coins found would be placed on the y axis while the year of coinage, from oldest to newest, would mark the x axis). The mean and median would be pulled toward these smaller values, but the median less dramatically so.

102. Trends Q5 ( )

Answer: Graph B. The graph of which product shows the smallest range?

Range is defined as the smallest value subtracted from the largest value:

Product A has a range of 80 – 20 = 60 years.
Product B has a range of 70 – 30 = 40 years.
Product C has a range of 80 – 20 = 60 years.
Product D also has a range of 80 – 20 = 60 years.

103. Trends Q6 ()

Answer: Graph B. The graph of which product shows the lowest standard deviation?

Product B has the lowest standard deviation because the values are most bunched around its arithmetic mean. In short, a low standard deviation tells us that the values are bunched around the arithmetic mean; a high standard deviation tells us that the values are spread out. Product A likely has the greatest standard deviation because the values are most spread out on both sides. Graphs C and D depict data that is fairly bunched around the mean.

104. Statistician ()

Choice D. Statement I is false. The difference between these two numbers (or values) may or may not be deemed important. Laypersons understandably but mistakenly associate the word "significant" with the word "importance." Statisticians use the word "significant" to mean "probable," so the words "statistically significant" translate as "statistically probable."

Statement II is true. The difference between two numbers (or values) may be determined to be statistically significant or not statistically significant. These two determinations are mutually exclusive. It's one or the other, and never both at the same time.

Statement III is true. The statistical odds that the difference between these two numbers or values cannot be relied upon are less than a 5% chance. A 5% chance is the same as a 1 in 20 chance. The words "cannot be relied upon" mean the same as "cannot be believed to be true."

Statement IV is true. There is at least a 95% chance that the difference between these two numbers or values can be relied upon and less than a 5% chance that this difference is due to chance. Voila!—We have cited the very definition of statistical significance. Of course, it expresses a definition both in terms of what is true as well as what is not true. That is, if we have a 95% or greater chance of being right, we also have a less than 5% chance of being wrong. Finally, "relied upon" may be interpreted as "probably true or probably real." So statement IV could have stated: "There is at least a 95% chance that the difference between these two numbers or values is probably true, and a less than 5% chance that this difference is not probably true (or not probably real)."

Appendix II – The World of Numbers

105. Fractions to Percents ( )

$\frac{1}{3} = 33\frac{1}{3}\%$ $\frac{2}{3} = 66\frac{2}{3}\%$ $\frac{1}{6} = 16\frac{2}{3}\%$ $\frac{5}{6} = 83\frac{1}{3}\%$

$\frac{1}{8} = 12.5\%$ $\frac{3}{8} = 37.5\%$ $\frac{5}{8} = 62.5\%$ $\frac{7}{8} = 87.5\%$

$\frac{1}{9} = 11.11\%$ $\frac{5}{9} = 55.55\%$

106. Decimals to Fractions ( )

The answer to the first question is found by adding 1 to the fractional equivalent of 0.2.

$$0.2 = \frac{1}{5} \qquad \text{Thus, } 1.2 = 1 + \frac{1}{5} = 1\frac{1}{5} = \frac{6}{5}$$

The answer to the second question is found by adding 1 to the fractional equivalent of 0.25.

$$0.25 = \frac{1}{4} \qquad \text{Thus, } 1.25 = 1 + \frac{1}{4} = 1\frac{1}{4} = \frac{5}{4}$$

The answer to the third question is found by adding 1 to the fractional equivalent of 0.33.

$$0.33 = \frac{1}{3} \qquad \text{Thus, } 1.33 = 1 + \frac{1}{3} = 1\frac{1}{3} = \frac{4}{3}$$

Each of the following three "fraction" problems equals 1!

$$\frac{5}{4} \times \frac{5}{10} \times \frac{8}{10} \times \frac{2}{1} = \frac{400}{400} = 1$$

$$\frac{\frac{3}{4} \times \frac{5}{6}}{\frac{5}{8}} = \frac{\frac{15}{24}}{\frac{5}{8}} = \frac{15}{24} \times \frac{8}{5} = \frac{120}{120} = 1$$

$$\frac{\frac{2}{9}}{\frac{1}{3} \times \frac{2}{3}} = \frac{\frac{2}{9}}{\frac{2}{9}} = \frac{2}{9} \times \frac{9}{2} = \frac{18}{18} = 1$$

107. Common Square Roots ( )

The answer to this square root problem is III, II & I. These statements are, in fact, already ordered from smallest to largest value, but the question requires ordering the values from largest to smallest.

I. $1 + 2.2 = 3.2$
II. $2 + 1.7 = 3.7$
III. $3 + 1.4 = 4.4$

Appendix III – Translating Words into Numbers

108. Leo ( )

Choice B. In cases where an age problem states, "E is 10 years older than L," the math translation is $E = L + 10$, not $E + 10 = L$. In cases where an age problem states, "5 years from now...," remember to add 5 years to both sides of the equation because both individuals (or in this case animals) will have aged 5 years.

First Equation:

Elmer is currently 10 year older than Leo.

$E = L + 10$

Second Equation:

In five years from now, Leo will be twice as old as Elmer.

$E + 5 = 2(L + 5)$
or:
$E + 5 = 2L + 10$

Next, we substitute for the variable E in order to solve for L. That is, substitute $L + 10$ (First Equation) for the variable E (Second Equation).

$L + 10 + 5 = 2L + 10$
$L + 15 = 2L + 10$
$L - 2L = 10 - 15$
$-L = -5$
$(-1)(-L) = (-1)(-5)$ [multiply through by -1]
$L = 5$

Therefore, Leo, the circus Lion, is 5 years old. Note that this also means that Elmer, the circus Elephant, is 15 years old. This calculation is derived from the first equation, $E = L + 10$.

109. Brad and Janet (🌶🌶)

Choice C. In cases where an age problem states, "A is twice as old as B," the math translation is $A = 2B$, not $2A = B$. In cases where an age problem states, "B is half as old as A," the math translation is $B = \frac{1}{2}A$, not $\frac{1}{2}B = A$.

First Equation:

> If Janet were eight years younger, she would be half as old as Brad is now.
>
> $J - 8 = \frac{1}{2}B$
>
> *or:*
>
> $2J - 16 = B$ [multiply each term by the number 2]

Second Equation:

> If Janet were two years older, she would be the same age as Brad is now.
>
> $J + 2 = B$

Next, we substitute for the variable B in order to solve for J. That is, we substitute for the variable B (in the first equation) with the expression $J + 2$ (in the second equation).

$$2J - 16 = J + 2$$
$$2J - J = 2 + 16$$
$$J = 18$$

Therefore, Janet is 18 years old. Note that this also means that Brad is 20 years old. This calculation can be derived from either the first or second equation, but it is easiest derived from the second equation, $J + 2 = B$.

110. Hardware ( )

Choice A. This problem further highlights a commonly entered scenario when translating words into math. If we have twice as many pencils as crayons, the algebraic expression is $P = 2C$, not $2P = C$.

The total weight of 2 hammers and 3 wrenches is one-third that of 8 hammers and 5 wrenches:

$$3(2H + 3W) = 8H + 5W$$

$$6H + 9W = 8H + 5W$$

$$4W = 2H$$

$$W = \frac{2}{4}H$$

$$W = \frac{1}{2}H$$

Choosing the correct answer is perhaps the trickiest step because the mathematical result may seem counterintuitive and may be interpreted in reverse. Since $W = \frac{1}{2}H$ or $2W = H$, this means that two wrenches are as heavy as one hammer. Stated another way, a single wrench is half the weight of one hammer.

111. Snooker ( )

Choice D. The classic way to solve an algebraic word problem is to identify two distinct equations and then substitute for one of the variables.

Set-up: Let x represent the number of general admission tickets and y represent the number of VIP admission tickets.

First equation:

$$x + y = 320$$
$$y = 320 - x \qquad \text{[Solve for } y \text{ in terms of } x.]$$

Second equation:

$$\$15x + \$45y = \$7{,}500$$
$$\$x + \$3y = \$500 \qquad \text{[Simplify by dividing each term by the number 15.]}$$

"Put the first equation into the second equation." That is, we substitute for the variable y in the first equation and solve for x.

$$\$x + \$3(320 - x) = \$500$$
$$\$x + \$960 - 3x = \$500$$
$$-\$2x = \$460$$
$$(-1)(-\$2x) = (-1)(-\$460)$$
$$\$2x = \$460$$
$$x = \frac{\$460}{\$2}$$
$$x = 230 \text{ tickets}$$

Therefore, 230 general admission tickets were sold. Note also that 90 VIP tickets were sold. We can make this calculation using either the first or second equation (but more easily the first equation):

$$230 + y = 320$$
$$y = 90$$

112. Chili Paste ()

Choice D. There are two basic approaches for solving this problem—solve for one variable using a single equation or solve for two variables using two equations. Scenarios 1, 2, and 3 use a single variable; scenario 4 uses a classic two-variable, two equations approach.

Scenario 1: $15S = 25(S - 40)$ [where S = small cans]

$S = 100$ small cans

Scenario 2: $25L = 15(L + 40)$ [where L = large cans]

$L = 60$ large cans

Therefore, $S = 100$ small cans

Scenario 3: $\frac{x}{15} - \frac{x}{25} = 40$ [where x = weekly needs of restaurant]

$x = 1{,}500$ ounces

Thus, $1{,}500 \div 15 = 100$ small cans

Scenario 4: The second equation in this scenario is tricky because it utilizes a fractional relationship involving the volume of small cans to large cans, that is, $\frac{15}{25}$ ounces. Let S represent the number of small cans and L represent the number of large cans.

i) $S = L + 40$

ii) $L = \frac{15}{25}S$ (or $S = \frac{25}{15}L$)

Substituting equation ii) into equation i) above:

Thus, $S = \frac{15}{25}S + 40$

$S = 100$ small cans

NOTE *Chili Paste* is an interesting but tricky problem and one which highlights multiple solutions. That's what's really interesting about math—"you can get downtown by taking many different roads." In other words, there are often multiple approaches to use to arrive at one single correct answer.

113. Bullet Train #1 ()

Choice B. The best way to approach this problem is in four steps: i) identify the correct formula, in this case $D = R \times T$, ii) fill in known values for R and T, iii) tackle the conversion from minutes to hours, and iv) solve the problem.

i) $D = R \times T$

ii) $D = \dfrac{d \text{ miles}}{q \text{ minutes}} \times x \text{ hours}$

iii) $D = \dfrac{d \text{ miles}}{q \cancel{\text{ minutes}}} \times x \cancel{\text{ hours}} \times \dfrac{60 \cancel{\text{ minutes}}}{1 \cancel{\text{ hour}}}$

iv) $D = \dfrac{60dx}{q} \text{ miles}$

114. Bullet Train #2 ()

Choice C. When solving for R, and converting from minutes to hours, we anticipate multiplying by 60.

i) $D = R \times T$

ii) $d \text{ miles} = R \times q \text{ minutes}$

$R = \dfrac{d \text{ miles}}{q \text{ minutes}}$

iii) $R = \dfrac{d \text{ miles}}{q \cancel{\text{ minutes}}} \times \dfrac{60 \cancel{\text{ minutes}}}{1 \cancel{\text{ hour}}}$

iv) $R = \dfrac{60d}{q}$ miles per hour

115. Bullet Train #3 ( )

Choice D. When solving for *T*, and converting from minutes to hours, we anticipate dividing by 60.

i) $D = R \times T$

ii) $d \text{ miles} = \dfrac{z \text{ miles}}{\text{minute}} \times T$

$$T = \frac{d \text{ miles}}{\dfrac{z \text{ miles}}{\text{minute}}}$$

iii) $T = d \cancel{\text{ miles}} \times \dfrac{\cancel{\text{minute}}}{z \cancel{\text{ miles}}} \times \dfrac{1 \text{ hour}}{60 \cancel{\text{ minutes}}}$

iv) $T = \dfrac{d}{60z}$ hours

116. Run-Run ( )

Choice B. The trickiest part of this problem is finding a proper rate (or speed) expressed algebraically. Rate is defined as a certain distance divided by a certain time frame. Here, the proper rate is $\frac{y}{9}$ not $\frac{9}{y}$ because distance is the numerator and time is the denominator. The problem is asking for time, where $T = \frac{D}{R}$.

i) $D = R \times T$

ii) $x \text{ yards} = \dfrac{y \text{ yards}}{9 \text{ seconds}} \times T$

$$T = \frac{x \text{ yards}}{\dfrac{y \text{ yards}}{9 \text{ seconds}}}$$

iii) $T = x \cancel{\text{yards}} \times \dfrac{9 \cancel{\text{seconds}}}{y \cancel{\text{yards}}} \times \dfrac{1 \text{ minute}}{60 \cancel{\text{seconds}}}$

iv) $T = \dfrac{9x}{60y} \text{ minutes}$

NOTE ✍ The following summary provides another, albeit more complicated, way to view the conversions made throughout this section. "When solving for distance or rate, we *multiply* by 60 when converting from minutes to hours (or seconds to minutes) and *divide* by 60 when converting from hours to minutes (or minutes to seconds). When solving for time, however, we *divide* by 60 when converting from minutes to hours (or seconds to minutes) and *multiply* by 60 when converting from hours to minutes (or minutes to seconds)."

Appendix IV – Tackling Squares and Cubes

117. Bigger or Smaller ( )

Choice E. This question examines the full gamut of possibilities with respect to the squaring and cubing of positive and negative numbers.

Let's take each answer choice.

Choice A. "All positive numbers, when squared, get bigger." This statement is not true. Positive fractions between 0 and 1, when squared, get smaller. Also, the positive number 1 stays the same when squared so it doesn't become bigger.

Choice B. "All positive numbers, when cubed, get bigger." This statement is not true. Positive fractions between 0 and 1, when cubed, get smaller. Also, the positive number 1 stays the same when squared so it doesn't become bigger.

Choice C. "Some negative numbers, when squared, stay the same." This statement is not true. All negative numbers when squared get bigger.

Choice D. "All negative numbers, when cubed, get smaller." This statement is not true. Negative fractions between 0 and –1, when squared, get bigger.

Choice E. "All negative numbers, when squared, get bigger." This statement is true. Negative numbers when squared become positive numbers and so they always become bigger.

118. Plus-Zero (🌶🌶)

Choice D. Always consider which of the "big" seven numbers —that is, 2, −2, 1, −1, $\frac{1}{2}$, $-\frac{1}{2}$, and 0—satisfy each of the conditions presented in statements I through III. When a problem states $x > 0$, three numbers should immediately come to mind: 2, 1, and $\frac{1}{2}$.

Statement I:

Could x^3 be greater than x^2?
Answer—yes.

Example $2^3 > 2^2$
Result $8 > 4$

Statement II:

Could x^2 be equal to x?
Answer—yes.

Example $1^2 = 1$
Result $1 = 1$

Statement III:

Could x^2 be greater than x^3?
Answer—yes.

Example $\left(\frac{1}{2}\right)^2 > \left(\frac{1}{2}\right)^3$

Result $\frac{1}{4} > \frac{1}{8}$

119. Sub-Zero ( )

Choice B. When a problem states $x < 0$, three numbers should immediately come to mind: -2, -1, and $-\frac{1}{2}$.

Statement I:

Is x^2 greater than 0?

Answer—absolutely. As long as x is negative, it will, when squared, become positive.

Statement II:

Is $x - 2x$ greater than 0?

Answer—absolutely. As long as x is negative the expression "$x - 2x$" will be greater than zero.

Statement III:

Is $x^3 + x^2$ less than 0?

Answer—not necessarily.

Example $\left(-\frac{1}{2}\right)^3 + \left(-\frac{1}{2}\right)^2$

$$-\frac{1}{8} + \frac{1}{4} = \frac{1}{8}$$

Result $\frac{1}{8} > 0$

Appendix V – Understanding How Variables Change

120. Quiz on How Variables Change (🌶🌶)

NOTE ✍ One tip that will likely prove very useful is to assume at the outset that variables *A, b,* and *c* are all equal to 1. Then when one variable is, say, doubled or halved, then substitute the numbers 2 or $\frac{1}{2}$ into the equation at hand and see how the other variable in question changes in value. Compare the resultant value to the original value of 1. In the particular case of Formula 2, assume initially that *b* and *c* are equal to 1 and the that *A* is equal to $\frac{1}{2}$.

1. Choice A. If *b* is doubled, then *A* will be doubled.

   The variable *b* is in the numerator (top) of the fraction. It therefore varies directly with variable *A*. When *b* increases, *A* increases; when *b* decreases, *A* decreases.

2. Choice D. If *b* is halved, then *A* will be halved.

   Again, variable *b* varies directly with variable *A*. When *b* decreases, *A* decreases; when *b* increases, *A* increases.

3. Choice D. If *c* is doubled, then *A* will be halved.

   The variable *c* varies indirectly with variable *A*. When *c* increases, *A* decreases; when *c* decreases, *A* increases.

4. Choice A. If *c* is halved, then *A* will be doubled.

   Again, the variable *c* varies indirectly with variable *A*. When *c* decreases, *A* increases; when *c* increases, *A* decreases.

5. Choice A. If *b* is doubled, then *A* will be doubled.

   Again, variable *b* varies directly with variable *A*. When *b* increases, *A* increases; when *b* decreases, *A* decreases.

6. Choice B. If *c* is tripled, then *A* will be tripled.

   Variable *c* varies directly with variable *A*. When *c* increases, *A* increases; when *c* decreases, *A* decreases.

7. Choice D. If *c* is halved, then *A* will be halved.

   Again, variable *c* varies directly with variable *A*. When *c* decreases, *A* decreases; when *c* increases, *A* increases.

8. Choice C. If *b* and *c* are both doubled, then *A* will be quadrupled.

   Variables *b* and *c* both vary directly with variable *A*. Thus, when either *b* or *c* increase, *A* increases; when either *b* or *c* decrease, *A* decreases.

9. Choice D. If *b* is doubled, then *A* will be halved.

   Variable *b* varies indirectly with variable *A*. When *b* increases, *A* decreases; when *b* decreases, *A* increases.

10. Choice A. If *c* is halved, then *A* will be doubled.

    Variable *c* also varies indirectly with variable *A*. When *c* decreases, *A* increases; when *c* increases, *A* decreases.

11. Choice C. If *b* is doubled, then *A* will be quadrupled.

    Variable *b* varies directly with variable *A*. If *b* is doubled, it will become four times its size. This will cause *A* to become four times its size.

    NOTE ☙ Variable *b* varies both directly and disproportionately with variable *A*. Since *b* is to the power of two, any value for *b* will be squared. Stated another way, variable *b* is squared and this means that an increase in *b* by a factor of 1 will lead to an increase in *A* by a factor of 4 (and vice versa).

12. Choice D. If *c* is doubled, then *A* will be halved.

    Variable *c* varies indirectly with variable *A*. When *c* increases, *A* decreases (and vice versa).

13. Choice A. If *b* is doubled, then *A* will be doubled.

    Variable *b* varies directly with variable *A*. When *b* increases, *A* also increases (and vice versa).

14. Choice C. If *c* is halved, then *A* will be quadrupled.

    Variable *c* varies indirectly with variable *A*. If *c* is halved, it will be reduced to one-quarter of its size. This will cause *A* to become four times its size.

    NOTE ☙ Variable *c* varies both indirectly and disproportionately with variable *A*. Since *c* is to the power of two, any value for *c* will be squared. In other words, variable *c* is squared and this means that a decrease in *c* by a factor of 1 will lead to an increase in *A* by a factor of 4 (and vice versa).

Answers to the Quiz

1. False. The percentage of people who are female at this conference is $\frac{1}{3}$ or $33\frac{1}{3}$%, not 50%.

   See Tip #9, *Chapter 1.*

2. False. For a given product, markup is always larger than margin.

   See Tip #18, *Chapter 3.*

3. False. The cost of the meal before tax and tip was \$100. Calculation: (\$132 ÷ 1.2) ÷ 1.1 = \$100.

   See Tip #11, *Chapter 1.*

4. False. Ratios tell us nothing about actual size or value; they tell us instead about relative size or value.

   See Tip #7, *Chapter 1.*

5. False. Multiplying a number by 1.2 is the same as dividing the number by the reciprocal of 1.2, which is 0.83, not 0.8.

   Case in point: \$100 × 1.2 = \$120; \$100 ÷ 0.83 = \$120.

   See Tip #12, *Chapter 1.*

6. False. Break-even occurs exactly where variable revenue (sales revenue less variable costs) equals total fixed costs.

   See Tip #21, *Chapter 3.*

7. False. The store item is now selling at a 44% discount, or 56% of its original price. For example, $100 less 20% equals $80, and $80 less 30% equals $56. A $44 discount on $100 is 44%. The trap here is that you can't add (or subtract) the percentages of different wholes.

   See Tip #2, *Chapter 1.*

8. False. Fifteen persons donated to both Charity A and Charity B.

   Charity A + Charity B – Both + Neither = Total
   $60 + 35 - x + 20 = 100$
   $x = 15$

   See Tip #14, *Chapter 2.*

9. False. If product A is selling for 20% more than product B, then the ratio of the selling price of product A to product B is 120% to 100%.

   See Tip #11, *Chapter 1.*

10. False. Data with a high standard deviation is spread out. Data with a low standard deviation is bunched.

    See Tip #29, *Chapter 4.*

Selected Bibliography

Dewdney, A.K. *200% of Nothing: An Eye-Opening Tour through the Twists and Turns of Math Abuse and Innumeracy.* New York: Wiley, 1993.

Graduate Management Admission Council. *The Official Guide for GMAT Quantitative Review.* 2nd ed. Hoboken, NJ: Wiley, 2009.

Green, Sharon Weiner and Ira K. Wolf. *Barron's GRE.* 17th ed. Hauppauge, NY: Barron's, 2007.

Julius, Edward H. *Rapid Math Tricks and Tips: 30 Days to Number Power.* New York: Wiley, 1992.

Lerner, Marcia. *Math Smart: Essential Math for These Numeric Times.* 2nd ed. New York: Princeton Review Publishing, 2001

"Math." *Math.com: The World of Math Online.* http://www.math.com/.

Nardi, William A. *How to Solve Algebra Word Problems.* 4th ed. New York: Arco, 2002.

OnlineMathLearning. http://www.onlinemathlearning.com/.

Paulos, John Allen. *Innumeracy: Mathematical Illiteracy and Its Consequences.* NewYork: Hill & Wang, 1988.

Stanton, Robert. *Math Power: Score Higher on the SAT, GRE, and Other Standardized Tests.* 3rd ed. NewYork: Kaplan Publishing, 2003.

Stapel, Elizabeth. *Purplemath.* http://www.purplemath.com/.

Stewart, Ian. *Professor Stewart's Cabinet of Mathematical Curiosities.* New York: Basic Books, 2009.

Wikipedia, The Free Encyclopedia, s.v. "Mathematics," http://en.wikipedia.org/wiki/Mathematics (accessed December 5, 2009).

Successmatics

cannot
– not
= can

Index

Numbers in *italics* (within brackets) indicate multiple-choice problems, 1 to 120. They are preceded by corresponding page numbers.

A mind is not a vessel to be filled,
but a fire to be kindled.
—Plutarch

About the Author

Brandon Royal (CPA, MBA) is an award-winning writer whose educational authorship includes *Power Thinking, Power Writing,* and *Power Grammar*. During his tenure working in Hong Kong for US-based Kaplan Educational Centers—a Washington Post subsidiary and the largest test-preparation organization in the world—Brandon honed his theories of teaching and education and developed a set of key learning "principles" to help define the basics of writing, grammar, math, and reasoning.

A Canadian by birth and graduate of the University of Chicago's Booth School of Business, his interest in writing began after completing writing courses at Harvard University. Since then he has authored a dozen books and reviews of his books have appeared in *Time Asia* magazine, *Publishers Weekly, Library Journal of America, Midwest Book Review, The Asian Review of Books, Choice Reviews Online, Asia Times Online,* and About.com.

Brandon is a five-time winner of the International Book Awards, a six-time gold medalist at the President's Book Awards, as well as recipient of the 2011 "Educational Book of the Year" award as presented by the Book Publishers Association of Alberta. He has also been a winner or finalist at the Ben Franklin Book Awards, the Global eBook Awards, the IPPY Awards, the USA Book News "Best Book Awards," and the *Foreword* magazine Book of the Year Awards.

To contact the author:
E-mail: contact@brandonroyal.com
Web site: www.brandonroyal.com

JAICO PUBLISHING HOUSE

Elevate Your Life. Transform Your World.

ESTABLISHED IN 1946, Jaico Publishing House is home to world-transforming authors such as Sri Sri Paramahansa Yogananda, Osho, The Dalai Lama, Sri Sri Ravi Shankar, Sadhguru, Robin Sharma, Deepak Chopra, Jack Canfield, Eknath Easwaran, Devdutt Pattanaik, Khushwant Singh, John Maxwell, Brian Tracy and Stephen Hawking.

Our late founder Mr. Jaman Shah first established Jaico as a book distribution company. Sensing that independence was around the corner, he aptly named his company Jaico ('Jai' means victory in Hindi). In order to service the significant demand for affordable books in a developing nation, Mr. Shah initiated Jaico's own publications. Jaico was India's first publisher of paperback books in the English language.

While self-help, religion and philosophy, mind/body/spirit, and business titles form the cornerstone of our non-fiction list, we publish an exciting range of travel, current affairs, biography, and popular science books as well. Our renewed focus on popular fiction is evident in our new titles by a host of fresh young talent from India and abroad. Jaico's recently established Translations Division translates selected English content into nine regional languages.

In addition to being a publisher and distributor of its own titles, Jaico is a major national distributor of books of leading international and Indian publishers. With its headquarters in Mumbai, Jaico has branches and sales offices in Ahmedabad, Bangalore, Bhopal, Chennai, Delhi, Hyderabad, Kolkata and Lucknow.

SINCE 1946